25 Springer Series in Solid-State Sciences

Edited by Manuel Cardona and Peter Fulde

Springer Series in Solid-State Sciences

Editors: M. Cardona P. Fulde H.-J. Queisser

Fundamental Physics of Amorphous Semiconductors

Proceedings of the Kyoto Summer Institute
Kyoto, Japan, September 8–11, 1980

Editor:
F. Yonezawa

With 91 Figures

Springer-Verlag Berlin Heidelberg New York 1981

Professor *Fumiko Yonezawa*
Research Institute for Fundamental Physics, Yukawa Hall, Kyoto Uni   ity
Kyoto 606, Japan

Series Editors:
Professor Dr. Manuel Cardona
Professor Dr. Peter Fulde
Professor Dr. Hans-Joachim Queisser
Max-Planck-Institut für Festkörperforschung, Heisenbergstrasse
D-7000 Stuttgart 80, Fed. Rep. of Germany

The Kyoto Summer Institute 1980 (KSI '80) was organized by Research Institute for
Fundamental Physics (RIFP), Kyoto University.

Organizing Committee

F. Yonezawa (RIFP; Chairwoman)	T. Matsubara (Kyoto Univ.)
Y. Nagaoka (RIFP)	M. Kikuchi (Sony)

Scientific Committee

K. Tanaka (ELH)	Y. Osaka (Hiroshima Univ.)
T. Shimizu (Kanazawa Univ.)	M. Okazaki (Tsukuba Univ.)
K. Morigaki (Univ. of Tokyo)	S. Minomura (Univ. of Tokyo)
	A. Yoshida (Nagoya Univ.)

Executive Committee

S. Nitta (Gifu Univ.)	Y. Hamakawa (Osaka Univ.)
M. Hirose (Hiroshima Univ.)	A. Hiraki (Osaka Univ.)
A. Ueda (Kyoto Univ.)	T. Ogawa (Tsukuba Univ.)

Publication Committee

F. Yonezawa (RIFP)	Y. Nagaoka (RIFP)
T. Ninomiya (Univ. of Tokyo)	T. Arai (Tsukuba Univ.)

Sponsored by Research Institute for Fundamental Physics, Kyoto University

Cosponsored by

Toa Nenryo Kogyo Co., Ltd.	Kyoto Ceramic Co., Ltd.
Anelva Corporation	Matsushita Electric Industrial Co., Ltd.,
Asahi Chemical Industry Co., Ltd.	Wireless Research Laboratory
CANON Research Center for New	Mitsubishi Electric Corporation, LSI
Technology & Future Products	Research and Development Laboratory
Fuji Electric Corporate Research and	Ricoh Co., Ltd.
Development, Ltd.	Sanyo Electric Co., Ltd., Research Center
Fuji Photo Film Co., Ltd.	Sharp Corporation
Hitachi Ltd., Central Laboratory	Teijin Ltd.
Komatsu Ltd.	Yamada Science Foundation

ISBN 3-540-10634-0 Springer-Verlag Berlin Heidelberg New York
ISBN 0-387-10634-0 Springer-Verlag New York Heidelberg Berlin

Offset printing: Beltz Offsetdruck, 6944 Hemsbach/Bergstr. Bookbinding: J. Schäffer oHG, 6718 Grünstadt.
2153/3130-543210

Preface

The Kyoto Summer Institute 1980 (KSI '80), devoted to "Fundamental Physics
of Amorphous Semiconductors", was held at Research Institute for Fundamental
Physics (RIFP), Kyoto University, from 8-11 September, 1980. The KSI '80 was
the successor of the preceding Institutes which were held in July 1978 on
"Particle Physics and Accelerator Projects" and in September 1979 on "Physics
of Low-Dimensional Systems". The KSI '80 was attended by 200 participants,
of which 36 were from abroad: Canada, France, Korea, Poland, U.K., U.S.A,
U.S.S.R., and the Federal Republic of Germany.

The KSI '80 was organized by RIFP and directed by the Amorphous Semiconductor group in Japan. A few years ago, we started to organize an international meeting on amorphous semiconductors as a satellite meeting of the
International Conference on "Physics of Semiconductors" held on September 1-5,
1980 in Kyoto. We later decided to hold the meeting in the form of the Kyoto
Summer Institute.

The Kyoto Summer Institute is aimed to be something between a school and
a conference. Accordingly, the object of the KSI '80 was to provide a series
of invited lectures and informal seminars on fundamental physics of amorphous
semiconductors. No contributed paper was accepted, but seminars were open.
In this line, the KSI '80 was undoubtedly successful and our above-mentioned
object was fulfilled. We owe the success of the KSI '80 to the speakers, to
the participants, to the sponsors and to all those who worked for the KSI '80.

The KSI '80 was partially sponsored by Toa Nenryo Kogyo Co., Ltd., the
Yamada Science Foundation and other organizations listed on a separate page.
The Organizing Committee acknowledges gratefully the financial support of
these organizations.

Thanks are also due to Prof. Morrel H. Cohen for his useful advice and
enormous mental support, and to Mrs. C. Hikami for her invaluable assistance.

Kyoto, Japan   *F. Yonezawa*
December, 1980

Contents

W. Paul   D. Weaire   G. Lucovsky   I. Solomon   V. L. Bonch-Bruevich   M. H. Brodsky

K. Tanaka   H. Fritzsche   W. E. Spear   P. G. LeComber   D. Adler   F. Yonezawa

KSI Lecturers, as sketched by *Denis Weaire*, with due apologies to all the subjects!

What are Non-Crystalline Semiconductors

Hellmut Fritzsche

The James Franck Institute and Department of Physics,
The University of Chicago
Chicago, IL 60637, USA

1. Introduction

We are presently participating in the development of a fascinating new field
of solid-state physics. Although glasses have been known for over 10,000
years and during the course of a day you are likely to encounter more non-
crystalline than crystalline solids, there is hardly a solid-state textbook
which mentions glassy or amorphous materials. The beginning of our field is
difficult to pin down. In the 1950's KOLOMIETS [1] showed that chalcogenide
glasses behaved like intrinsic semiconductors and that their electrical conductivity could not be increased by adding dopants. In 1957 SPEAR [2]
reported the first drift mobility measurements in vitreous Se and TAUC [3]
reported the first studies on amorphous Ge in the early 1960's. Many more
scientists were drawn to our field through the work of OVSHINSKY [4]: the
discovery of switching and memory effects in chalcogenide glasses. These
as well as optical memory effects, imaging, photodoping, and the reversible
photostructural changes suggested possibilities for new applications of non-
crystalline semiconductors. These phenomena demonstrated that there was a
large field of material science which was virtually unexplored.

The scope of our field was further widened by the discovery of SPEAR and
LeCOMBER a few years ago [5] that glow-discharge-deposited amorphous Si can
be doped both n-type and p-type. This material contains a sufficiently low
concentration of defects that it is now the most interesting prototype
amorphous semiconductor. Furthermore, it promises application for cheap and
large area photovoltaic and photothermal devices.

In this paper I shall first discuss some important concepts which clarify
the fundamental difference between the two main classes of noncrystalline
semiconductors, the glasses and the amorphous films. In this discussion I
shall call a glass only those materials which can be quenched from the supercooled melt and usually exhibit a glass transition. The term amorphous will
be restricted to noncrystalline materials which can normally be prepared only
in form of thin films by deposition on substrates which are kept sufficiently
cool to prevent crystallization. In the remainder of the paper I shall employ
these ideas in a survey of some of the major problems confronting us in the
study of noncrystalline semiconductors.

2. The Homogeneous Random Covalent Network

The atoms are completely disordered in a gas but not in noncrystalline semiconductors. Here the chemical nature of the atoms dictates the directed covalent bonding arrangements to their nearest neighbors. The first model of an
ideal glass, shown in Fig. 1a, was proposed by ZACHARIESEN [6] in 1932; the
figure is a two-dimensional representation of an A_2B_3 glass such as As_2S_3.

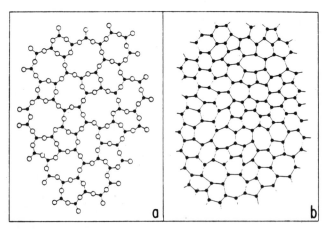

Fig.1a,b Two-dimensional continuous random network. (a) ZACHARIASEN's diagram for an A_2B_3 glass, (b) threefold-coordinated elemental random network

A definite short range order is imposed as each atom fulfills its chemical valency requirement according to the 8-N rule. Stronger heteropolar bonds (As-S) are favored over homopolar bonds (S-S,As-As) and the bond lengths are within 1% the same as those found in crystals. The figure illustrates that there are even and odd membered rings of different sizes. This type of disorder affects the electronic states near the band edges as explained by YONOZAWA later [7]. Another major element of randomness is the variation in bond angles. There we come to a very important point which leads to a differentiation between glasses and amorphous materials: the flexibility of covalent bond angles is largest for the two-fold coordinated group VI elements and least for the tetrahedrally coordinated group IV elements. The reason for this is the greater variety of admixture from other atomic orbitals to the covalent bond when the coordination number is less than the number of valence electrons. In SiO_2 glasses the oxygen atoms bridging the Si tetrahedra provide the essential flexibility which is needed to form a random covalent network without much strain. If one attempts to form a covalent random network without the flexing bridges of group VI elements one obtains amorphous Si which is highly overconstrained and no longer a glass. In Fig. 1b this step is illustrated by showing amorphous As because the structure of glassy SiO_2 and of amorphous Si are not easily sketched in two dimensions.

The concept of a homogeneous random network has been extensively used in theoretical studies for both amorphous and glassy semiconductors [8-10], but it fails to encompass the fundamental difference between the two classes of materials. This difference, which is essentially due to the mismatch between bonding constraints and the number of degrees of freedom in three dimensions, and to the flexibility required to accomodate the mismatch, has been treated quantitatively the PHILLIPS [11] as follows.

Consider a binary alloy A_xB_{1-x} with only the short-range bonding interactions. Further let us introduce the average coordination number m defined by

$$m = xN_{cn}(A) + (1-x) N_{Cn}(B) \tag{1}$$

An empirical justification for this procedure will be presented later. The number of constraints N_{CO} per atom is then given by

$$N_{CO} (m) = m/2 + m (m-1) /2 \qquad\qquad (2)$$

$$= m^2/2 \qquad\qquad (3)$$

The first term on the right hand side of (2) is given by the bond stretching interaction and the second term by the bond-bending interactions which are assumed for simplicity to be equal for the A and B atoms. Equating the number of constraints to the three spatial degrees of freedom, $N_{CO} = 3$, PHILLIPS obtains an average coordination corresponding to an optimal connectivity [11],

$$m_c = \sqrt{6} = 2.45 \qquad\qquad (4)$$

Hence the glass-forming tendency is greatest when the short-range order imposed by bond stretching and bending forces is just sufficient to exhaust the local degrees of freedom. In Fig.2 we sketch a classification of non-crystalline solids based on these concepts. The average coordination m is shown decreasing from left to right because of the familiar arrangements of atoms in the periodic table. On the left-hand side of m_c the internal strain increases with m; toward the right, the entropy increases with decreasing m because the materials become insufficiently crosslinked. Glasses are normally restricted to $3 > m \geq 2$. Materials with higher connectivity $4 \geq m \geq 3$ are over-constrained amorphous. Those having lower connectivity $m < 2$ are undercross-linked amorphous. Examples of this latter group are amorphous films of I_2 and Br_2, presently being studied by LANNIN [12], and amorphous films of inert gasses. The mean coordination m = 4 separates noncrystalline metals from semiconductors or insulators.

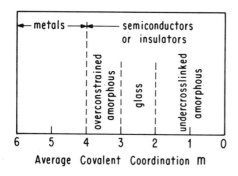

Fig.2 Classification of non-crystalline solids

This classification suffers as any other from uncertainties around the borderlines, near m = 3 and m = 2 for instance, since long range interactions, ionicity and size effects are neglected in this simple picture. Moreover, an interesting exception to the rule set forth by PHILLIPS is the class of materials known as tetrahedral glasses [13], which are obtained by quenching the melt of compounds such as $CdAs_2$, $CdGeAs_2$, and $CdGeP_2$. These have average coordinations m = 4. It appears that $A^{II}C_2^{V}$ is the essential building unit of these glasses since the concentration of group IV elements can be varied over wide limits.

3

I discussed the homogeneous random covalent network to emphasize the prevalence of short-range order and the distinction between amorphous and glassy materials. Strong doubts are presently being raised regarding the homogeneity of non-crystalline materials.

3. Topological and Compositional Heterogeneities

Even glasses with average coordination $m \simeq m_c$ contain a considerable strain energy. A measure of this energy is the difference in heats of solution between the glass and the corresponding crystal or the enthalpy of crystallization. 2.16 Kcal/mol are found for example for the enthalpy difference between vitreous silica ($m = 2.67$) and α-quartz, and 1.52 Kcal/mol when compared with α-cristobalite [14]. The origin of this energy is to be found not only in the variations in bond angles and to a smaller degree in bond lengths, but in accumulation of strain of longer range forces and van der Waals interactions associated with lone-pair electrons of Group V and VI elements in the bond-free directions. This strain energy can be uniformly distributed, but a lower energy state can also be established by forming low-strain grains or islands or strain relieving voids. Thus a network of voids is found in the highly overconstrained amorphous films of evaporated Si and Ge, as evidenced by the large density deficit (10-15%) of these materials with respect to their crystals and by small angle X-ray scattering [15], electron microscopy [16], and by porosity studies [17].

Another example of heterogeneity is the hydrogen-rich interconnecting tissue in the granular structure of glow-discharge deposited a-Si:H alloys [18]; both columnar and pebble-like morphologies are easily seen on electronmicrographs for certain growth conditions [19], as shown in Fig.3. NMR studies reveal compositional heterogeneities even in a-Si:H films which appear homogeneous under the electronmicroscope [18]. A further example of large-size domain formation in a 600 Å thick evaporated film of As_2Se_3 is shown in Fig.4. This transmission electronmicrograph [20] shows hexagonal domains which have diameters of 1000 ± 100 Å and are separated by about 30 Å wide troughs which nearly penetrate through the film. Even though evaporated films do not have the fully polymerized and crosslinked structure of relaxed bulk glasses, those domains may possibly demonstrate the accumulation of strain energies from long range interactions and a subsequent strain-relief mechanism in such thin films [21]. Once such a domain network is formed during film growth, the different surface mobilities of the gas phase constituents will likely lead to a connecting tissue with a composition which is different from that in the island or grains. Heterogeneities of this kind are then difficult to eliminate by annealing.

We now turn our attention to structural units in $GeSe_2$ and As_2Se_3 type glasses of the scale 15-30Å, i.e. the medium-range order. The presence of a very sharp first peak in the diffraction spectra of these semiconducting glasses has been associated with polyatomic, cagelike clusters containing 8-20 atoms or more [11,22]. UEMURA et al. [23] reported that the first peak of neutron diffraction spectra of some materials even sharpens as they are heated above T_g and that it is still observed above the crystalline melting temperature. This as well as the presence in vitreous GeS_2 of an extra A_1 Raman line which is absent from the Raman spectra of the crystals rule out a microcrystallite theory [22]. One difficulty in modeling the structure of these glasses is the great flexibility in the choice of the molecular units. This greater freedom of choice may actually be the basis for the glass-forming tendency. The similarity of the radial distribution or intensity functions calculated from various model structures is another problem [24,25]. One might suggest to start with the random network model for As and insert a flexing S or Se atom between each As-As bond. However, such a model would not contain the layer-like molecular units which make up the crystals.

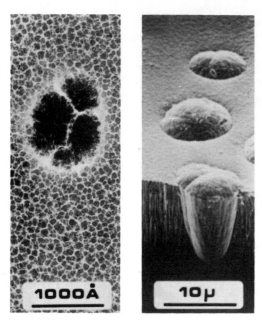

(A)   **(B)**

Fig.3 (a) Transmission electron micrograph of nodular feature in a-Si:H film prepared at 250°C with 25W power; (b) scanning electron micrograph of columnar growth features in a-Si:H film prepared at 230°C with 25W power. Courtesy of KNIGHTS [19]

Correlating the present data not only from diffraction but from infrared and Raman experiments, PHILLIPS suggests the presence of large molecular units shaped like outrigger rafts of various conformations [11,22]. At this point I can only draw attention to this ongoing work which attempts to describe the medium-range order that appears to exist not only in the glass but also above the glass transition temperature.

A strong medium-range order of a different kind is found in As_2S_3 and As_2Se_3 films deposited by evaporation [26-28]. This is illustrated in Fig.5 which compares Raman spectra of the bulk glass with those of an evaporated film before and after annealing [29]. Here the molecular units in the freshly deposited film derive from those present in the gas phase. Subsequent annealing produced polymerization toward the structure of the bulk glass and a decrease in the intensity and sharpness of the first diffraction peak.

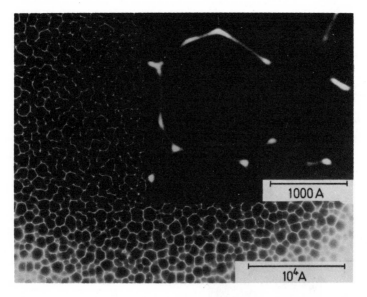

Fig.4 Transmission electron micrograph of domain wall network in a 600Å thick evaporated film of glassy As2Se3. The bright lines are deep troughs which are nearly normal to the plane of the film. Courtesy of CHEN et al. [20]

   In this section I gave a few examples of the growing evidence for medium-range structure and heterogeneities in both amorphous and glassy noncrystal - line materials. In the following I mention some basic differences between glasses and amorphous materials.

4. Glass Formation and Amorphous States

The melts of most crystalline solids have viscosities of less than 1 poise near the melting temperature T_m. The viscosity of water, in comparison, is about 0.1 poise. Some of these melts can be supercooled by perhaps 10-20 degrees below Tm but the presence of impurities or a slight disturbance will lead to rapid crystallization. A low viscosity facilitates the diffusion of atoms or molecules which is needed to transform the liquid into a crystalline solid. The melts of glass-forming materials behave entirely differently. Their viscosity increases rapidly with decreasing temperature [30] according to

$$\eta = \eta_o \, exp\left[-A/(T-T_o)\right]$$

(5)

and reaches a value of order 10^7 poise near T_m. Nucleation and growth of crystals becomes difficult and the supercooled liquid remains stable over several hundred degrees. Its viscosity continues to increase with decreasing T. Near $\eta = 10^{14}$ poise the supercooled liquid becomes a solid glass. The large increase in viscosity indicates that the structural units, mentioned in the previous section, are already formed during supercooling.

6

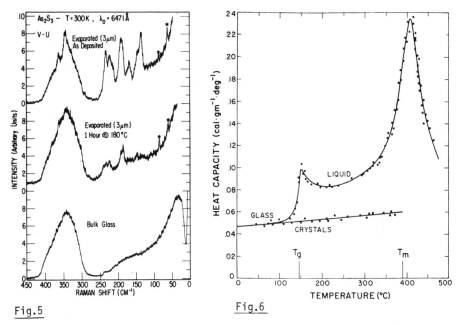

Fig.5

Fig.6

Fig.5 Raman spectra of (a) an evoporated vitreous As_2S_3 film, (b) after
annealing at 180°C for 1 hour, (c) bulk glass. Asterics indicate instrumental
artifacts. Courtesy of SOLIN et al. [29]

Fig.6 Heat capacity of glassy, liquid, and crystalline $Ge_{17}Te_{83}$ as a function of temperature. Courtesy of DeNEUFVILLE [33]

    The process of glass transition which leads to the freezing of a super-
cooled liquid at T=Tg into a noncrystalline solid (glass) is still an actively
pursued theoretical problem [31,32]. Empirically the glass transition temperature can be measured with an accuracy of about one percent or a few degrees.
As a glass is heated, abrupt changes in thermal expansion, compressibility,
specific heat and other properties occur at the glass transition temperature
T_g. This is illustrated in Fig. 6 for $Ge_{17}Te_{83}$, which is the eutectic
composition of the GeTe-Te system [33]. At low T the heat capacity c_V of the
glass and of the polycrystalline eutectic are equal to that of a harmonic
oscillator solid. At T_g the glass becomes liquid. The additional motional
degrees of freedom available in the liquid cause the sudden increase in c_V
at T_g. Above T_g the structure of the supercooled liquid resembles that of
the glass. As the temperature is further raised the structure of the liquid
may change in a complicated manner. In the example shown in Fig.6 one
observes a kind of order-disorder transition between 300 and 500K. The area
under c_V peak centered at the melting temperature T_m of the eutectic is
smaller than the heat of the fusion of pure Te. This suggests that vestiges
of short range order still remain above T_m which may disappear when T is
raised further [33]. I might remark here that metallic glasses do not show
the characteristic increase in viscosity. Their melts remain very fluid,
very high quench rates are needed to produce the non-crystalline solid, and
glass transitions are very seldom observed.

7

An experimental justification for using the average coordination number m in characterizing the connectivity of semiconducting glasses is demonstrated in Fig.7 which shows the relationship between the optical gap E_0 and the glass transition temperature T_g. Materials, including elements, binaries and ternaries, but having the same mean coordination m, lie essentially on the same line; the other important parameter which distinguishes them is the average strength of the covalent bonds, which increases with increasing E_0.

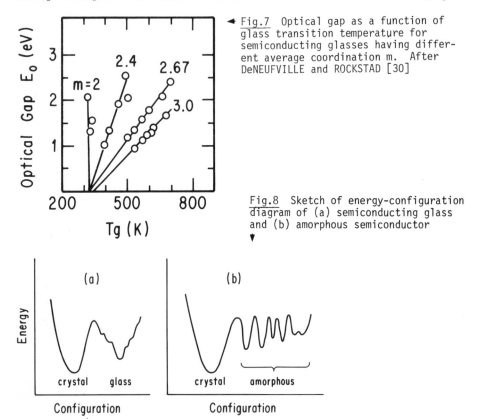

Fig.7 Optical gap as a function of glass transition temperature for semiconducting glasses having different average coordination m. After DeNEUFVILLE and ROCKSTAD [30]

Fig.8 Sketch of energy-configuration diagram of (a) semiconducting glass and (b) amorphous semiconductor

A major theoretical difficulty in describing the conditions for and the kinetics of the transition between a supercooled liquid and a glass is that the glass is not a thermodynamical equilibrium state. Nevertheless the macroscopic properties of a large number of glasses are very reproducible. Unstable glasses which require extreme quench rates are exceptions and are excluded from this bird-eye-view discussion. This empirical evidence suggests that the energy depends on a general configurational coordinate as sketched in Fig.8a. The energy minimum of the glass lies higher than the minimum which corresponds to the crystalline state because of the strain energies mentioned earlier. The barrier between the two main minima represent the kinetic hindrance which prevents crystallization of the glass below T_g. Substantial atomic rearrangements are necessary to accomplish this. The subsidiary minima on the glass side of the sketch indicate less stable higher energy configurations for instance of evaporated glass films which can be converted to the minimum energy glass configuration by annealing near T_g.

The somewhat wild sketch of Fig.8b is intended to illustrate the situation for amorphous films. These can exist in many different configurations depending on the preparation techniques and for each technique on a variety of preparation and substrate parameters. In contrast to glasses, the amorphous films can exist in many non-crystalline configurational states. Annealing can lower the energy of a given amorphous film. However, it cannot transform the configuration of an overconstrained amorphous film from one state to another. Major atomic rearrangements would be required to accomplish this which lead instead to crystallization. This figure is meant to reflect our experience that even annealed amorphous films prepared in various ways are different. An additional cause for the observed differences are of course impurities which are unintentionally incorporated during growth. As discussed above, the minimum energy of the homogeneous random covalent network might lie higher than those of strain-relieved heterogeneous structures. I hope Fig.8b does not convey a too pessimistic point of view regarding our ability to gain an understanding of the properties of amorphous films in general. In the case of plasma-deposited a-Si:H film we are after all dealing with very exciting energy minima in configuration space. However, the decomposition of these alloys by effusion of hydrogen at elevated temperatures is an additional problem that is not considered in the sketch of Fig.8b.

5.   Defects in Non-Crystalline Materials

5.1   Chalcogenide Glasses

Thermodynamic arguments require that at finite temperatures all crystals contain defects [34]. This can be seen in a simple way. Let ΔG be the free energy for creating a point defect. The equilibrium density of such defects at T is then $N_A \exp(-\Delta G/kT)$ where the factor N_A (the number of lattice sites) arises from the entropy of mixing. The defect density in crystals normally does not continue to decrease to zero with decreasing temperature but freezes in near a temperature at which healing of such defects by diffusion and lattice reconstruction ceases.

By the same argument, all glasses have intrinsic defects defined as deviations from the lowest energy bonding arrangements [35,36]. At present we exclude from consideration defects arising from foreign atoms. The temperature at which a given equilibrium density of intrinsic defects freezes in is probably close to T_g. Of all possible deviations from the ideal bonding configuration, the one that requires the least amount of energy will naturally be present at the highest concentration. KASTNER et al. [35,36] proposed that the least-costly bonding deviations in chalcogenide materials can be achieved when two defects are always created at the same time: a positively charged overcoordinated atom and a negatively charged undercoordinated atom. These are called valence-alternation pairs (VAP), because the valence or coordination is altered for the defect atoms.

Since the total number of bonds in the glass has not changed by creating these defects two at a time, the energy needed for creation of a pair is relatively small. It is approximately the energy required to place the negative charge onto the under-coordinated atom (reduced by relaxation effects of the surrounding matrix). In typical chalcogenide glasses this results in a VAP defect density of about $10^{17} cm^{-3}$ in agreement with defect densities measures by photo-induced spin resonance and absorption and by photoluminescence [37].

These VAP centers have all the features of the defects postulated earlier by STREET and MOTT [38] and MOTT et al. [39] for explaining a large number of electronic properties of chalcogenide glasses. In particular, these defects are associated with a negative effective correlation energy, which means that the reaction

$$2D^0 \rightarrow D^+ + D^- \tag{6}$$

is exothermic and that the charged defect centers can interconvert by charge transfer into their oppositely charged counterpart:

$$e + D^+ \rightleftarrows D^0 \rightleftarrows D^- + h \tag{7}$$

Here D^0, D^+ and D^- denote the different charge states of a defect center. In the VAP model the defects can be a group V, VI or VIII atom; group IV atoms are ruled out because their coordination cannot exceed four in a covalent matrix. Hence, bonding constraints may hinder the interconversion expressed by (7) for some defects in glasses containing appreciable amounts of group IV atoms [40]. Heterogeneities in the glass of the kind discussed in section 3 may introduce variations in the density of VAP centers as their creation energy depend on the local strain and composition. The VAP defect centers are normally diamagnetic. Since the concentration of unpaired spins in relaxed glasses is less than $10^{15}cm^{-3}$ [41], it is not likely that defect centers of entirely different nature are present. Exposure of chalcogenides to light at low temperatures brings these centers, by capture of photo-excited charge carriers, into their neutral and paramagnetic state [37]. This allows a study of their atomic structure by electron spin-resonance [37,42]. BIEGELSEN and STREET [43] have shown that VAP centers can also be photo-created because the energy required is less than the optical gap energy. It is not yet certain whether this process accounts for the light-induced shift of the optical absorption edge known as photo-darkening, or only for part of it.

Above T_g the concentration of valence alternation defects increases exponentially with temperature. I expect that they play an important role in aiding the formation of the structural units in the supercooled melt which give rise to the increase and narrowing of the first sharp diffraction peak mentioned earlier.

5.2   Amorphous Films

Defects in amorphous semiconductors probably result from a strain-relief mechanism and from bonding misfits during the growth process. Here we restrict the discussion to overconstrained amorphous materials. Except for a-As, these materials contain group IV elements as major constituants which cannot undergo valence alternation. A variety of paramagnetic and diamagnetic defects have been proposed and discussed by ADLER [44] for a-Si and a-Si:H. The dominant paramagnetic center in these films has a g-value of 2.0055 and is attributed to a dangling bond. No evidence has been found for the presence of defects having a negative effect correlation energy, despite some early claims to the contrary. The overconstrained nature of amorphous films suggests that the defects might not be randomly distributed but could be predominantly located at internal voids and strain-relief interfaces between low-strain regions. The close correlation between the photoluminescence efficiency and the spin density in a-Si:H indicates, on the other hand, that the distribution of at least the paramagnetic defect centers is nearly random [45].

10

6.   Conclusions

In the past we have learned much about the electronic properties of non-crystalline semiconductors without a full knowledge of their atomic structure. However, it appears that future progress depends first on improved preparation techniques which enable us to avoid structural and compositional heterogeneities in amorphous films as well as on theories which are applicable to non-crystalline materials which contain medium-range and larger heterogeneities. Substantial deviations from a homogeneous random covalent network will affect the electronic transport properties most strongly. Phenomena and properties which are particularly poorly understood both in films and glasses include the ac conductivity, the photoconductivity, the Meyer-Neldel relation between the preexponential factor and the activation energy of the conductivity [46], the different activation energies derived from thermopower and conductivity measurements [47,48], and the anomalous sign of the Hall coefficient [49,50].

I suppose that several of these phenomena are strongly influenced by heterogeneities. Consider for instance the extreme case of a semiconductor consisting of grains surrounded by thin interfacial material which is less conducting and which acts as potential barriers. Theoretical studies [51-55] as well as experiments on polycrystalline films [53,56-58] have shown that the Hall coefficient measures fairly accurately the carrier concentration in the grains. The activation energy of the conductivity on the other hand is roughly the sum of the carrier activation energy and the barrier height of the interfacial material. Hence the effective Hall mobility [50] contains the barrier height activation energy, which in turn may depend on temperature [58]. At low temperatures one expects that the carriers tunnel through the barrier.

In this model, the temperature dependence of the thermopower is governed essentially by the carrier activation energy in those grains which are in contact with the electrodes because one measures an open circuit voltage due to a temperature difference of the electrodes [59]. This activation energy is less than that of the conductivity. In many glassy and amorphous semiconductors one finds an energy difference between conductivity and thermopower measurements of about 0.15 eV [47,48] comparable to the activation energy of the Hall mobility [50]. Moreover, the common observation of the Meyer-Neldel relation [60] in heterogeneous semiconductors also suggests that one must be concerned about the possible presence of medium-range order and microstructure in non-crystalline semiconductors. Inhomogeneities will undoubtedly have a profound effect on the dielectric loss mechanisms and the ac conductivity [61,62]. I trust that some of these problems will be discussed in the following papers.

Acknowledgements

I gratefully acknowledge fruitful discussion with S.R.Ovshinsky, J.C.Phillips, and E.A.Schiff. The work was supported by NSF Grant No. DMR80-09225.

References

1.   B.T.Kolomiets, Proc. Intl. Conf. on Semiconductor Physics, Prague 1960, (Czechoslovak Academy of Sciences, 1961) p.884, Phys. Stat. Solidi 7, 359,713 (1964).
2.   J.Tauc, R. Grigorovici and A.Vancu, Phys. Stat. Solidi 15, 627 (1966).
3.   W.E.Spear, Proc. Phys. Soc. (London), 870, 1139 (1957); 76, 826 (1960).
4.   S.R.Ovshinsky, Phys. Rev. Lett. 21, 1450 (1968).

5. W.E.Spear and P.G.LeComber, Solid State Commun. 17, 1193 (1975).
6. W.H.Zachariasen, J.Am. Chem. Soc. 54, 3841 (1932).
7. F. Yonozawa, this volume.
8. D.E.Polk and D.S.Boudreaux, Phys. Rev. Lett. 31, 921 (1973.
9. G.N.Greaves and E.A.Davis, Phil. Mag. 29, 1201 (1974).
10. R.J.Bell and P. Dean, Phil. Mag. 25, 1381 (1972).
11. J.C.Phillips, J.Non-Cryst. Solids 34, 153 (1979).
12. J.S.Lannin and B.V.Shanabrook, Proc. 15th Intern. Conf. on the Physics of Semiconductors, Kyoto (1980).
13. L.Cervinka, A.Hruby, M.Matyas, T.Simecek, J.Skacha, L.Stourac, J.Tauc, V.Vorlicek, P.Höschel, J.Non-Cryst. Solids, 4, 258 (1970); M.Popescu, R.Manaila, and R.Grigorovici, J.Non-Cryst. Solids, 23, 229 (1977).
14. G.Scherer, P.J.Vergano, and D.R.Uhlmann, Phys. Chem. Glasses, 11, 53 (1970).
15. S.C.Moss and J.F.Graczyk, Phys. Rev. Lett. 23 , 1167 (1969).
16. A.Barna, P.B.Barna, G.Radnoczi, L.Toth and P.Thomas, Phys. Stat. Solidi (a) 41, 81 (1977.
17. H.Fritzsche and C.C.Tsai, Solar Energy Materials 1, 471 (1979).
18. J.A.Reimer, R.W.Vaughan and J.Knights, Phys. Rev. Lett. 44, 193 (1980).
19. J.Knights, J.Non-Cryst. Solids 35/36, 159 (1980); Japan J.Appl. Phys. 18, 101 (1979); E.A.Schiff, P.D.Persans, H.Fritzsche and V.Akopyan, Appl. Phys. Lett. (in press).
20. C.H.Chen, J.C.Phillips, K.L.Tai, and P.M.Bridenbaugh, (to be published).
21. J.C.Phillips, Phys. Rev. Lett. 42, 1151 (1979).
22. J.C.Phillips, Phys. Rev. B21, 5724 (1980); J.Non-Cryst. Solids 35/36, 1157 (1980); ibid (in press).
23. O.Uemura, Y.Sagara, D.Muno and T.Satow, J.Non-Cryst. Solids 30, 155 (1978).
24. P.H.Gaskell and I.D.Tarrant, Phil. Mag. B42, 265 (1980).
25. R.Grigorovici, J. Non-Cryst. Solids 35/36,1167 (1980).
26. A.J.Apling and A.J.Leadbetter, Amorphous and Liquid Semiconductors, ed. by J.Stuke and W.Brenig (Taylor and Francis, London, 1974)p.457.
27. A.J.Apling, A.J.Leadbetter, and A.C.Wright, J.Non-Cryst. Solids 23, 369 (1977).
28. J.P.deNeufville, R.Seguin, S.C.Moss and S.R.Ovshinsky, Amorphous and Liquid Semiconductors, ed. by J.Stuke and W.Brenig (Taylor and Francis, London, 1974) p.737; R.Nemanich, G.A.N.Connell, T.M.Hayes and R.A.Street, Phys. Rev. B18, 6900 (1978).
29. S.A.Solin and G.N.Papatheodorou, Phys. Rev. B15, 2084 (1977).
30. J.P.DeNeufville and H.K.Rockstad, Amorphous and Liquid Semiconductors, ed. by J.Stuke and W.Brenig (Taylor and Francis, London, 1974) p.419.
31. J.C.Phillips, J.Non-Cryst. Solids (in press).
32. M.H.Cohen and G.S.Grest, Phys. Rev. B20, 1077 (1979); G.S.Grest and M.H.Cohen, Phys. Rev. B21, 4113 (1980);(to be published).
33. J.P.deNeufville, J. Non-Cryst. Solids, 8-10, 85 (1972).
34. R.A.Swalin, Thermodynamics of Solids (J.Wiley, New Y ork, 1972) p.263.
35. M.Kastner, D.Adler, and H.Fritzsche, Phys. Rev. Lett. 37, 1504 (1976).
36. M.Kastner and H.Fritzsche, Phil. Mag. B37, 199 (1978).
37. S.G.Bishop, U.Strom, and P.C.Taylor, Phys. Rev. Lett. 34, 1346 (1975); 36, 543 (1976); Phys. Rev. B15, 2278 (1977).
38. R.A.Street and N.F.Mott, Phys. Rev. Lett. 35, 1293 (1975).
39. N.F.Mott, E.A.Davis, and R.A.Street, Phil. Mag. 32, 961 (1975).
40. R.A.Street and G.Lucovsky, Solid State Commun. 31, 289 (1979).
41. S.C.Agarwal, Phys. Rev. B7, 685 (1975).
42. P.Gaczi and H.Fritzsche (to be published).
43. D.K.Biegelsen and R.A.Street, Phys. Rev. Lett. 44, 803 (1980).
44. D.Adler, Phys. Rev. Lett. 41, 1755 (1978).

45. R.A.Street, J.C.Knights and D.K.Biegelsen, Phys. Rev. B18, 1880 (1978).
46. H.Fritzsche, Solar Energy Materials 3, 447 (1980).
47. P.Nagels, Amorphous Semiconductors, ed. by M.H.Brodsky, Topics in Appl. Physics Vol. 36 (Springer Verlag, New York 1979) p.113.
48. W.Beyer and H.Overhof, Solid State Commun. 31,1 (1979); W.Beyer, R. Rischer and H.Overhoff, Phil. Mag. B39, 205 (1979).
49. P.G.LeComber, D.I.Jones, and W.E.Spear, Phil. Mag. 35, 1173 (1977).
50. J.Dresner, Appl. Phys. Lett. 37, 742 (1980).
51. J.Volger, Phys. Rev. 79, 1023 (1950).
52. R.L.Petritz, Phys. Rev. 104, 1508 (1956).
53. H.Berger, Phys. Stat. Solidi 1, 739 (1961).
54. R.H.Bube, Appl. Phys. Lett. 13, 136 (1968).
55. K.Lipskis, A.Sakalas and J.Viscakas, Phys. Stat. Solidi (a) 4, K217 (1971).
56. R.Kassing and W.Bax, Japan J. Appl. Phys. Suppl. 2, 801 (1974).
57. John Y.W.Seto, J.Appl. Phys. 46, 5247 (1975).
58. J.W.Orton, B.J.Goldsmith, M.J.Powell and J.A.Chapman, Appl. Phys. Lett. 37, 557 (1980); and references quoted herein.
59. H.Fritzsche, (to be published).
60. W.Meyer and H.Neldel, Z. tech. Phys. 12, 588 (1937).
61. J.Volger, Progress in Semiconductors 4, 207 (1960).
62. L.K.H.VanBeek, Progr. Dielectrics 7, 69 (1967).

Defects in Covalent Amorphous Semiconductors

David Adler

Department of Electrical Engineering and Materials Science, and
Center for Materials Science and Engineering,
Massachusetts Institute of Technology
Cambridge, MA 02139, USA

1. Structure of Amorphous Solids

Before any analysis of the physical properties of an amorphous solid can be undertaken, a knowledge of the structure is essential. Of course, this is true of a crystalline solid as well. However, because of the lack of long-range periodicity in amorphous materials, it is much more difficult to determine the structure. In addition, the preparation methods necessary to obtain the material in the amorphous phase require the consideration of several issues that are not usually contemplated when investigating crystals. First and foremost of these is the question of composition: we must determine what is in the material. If the solid has been prepared as a bulk glass, there are not usually too many compositional surprises. However, many of the most important amorphous solids cannot be fabricated as bulk glasses but are generally deposited as thin films directly from the vapor phase, while even compositions which are relatively easy glass formers are often also deposited in thin-film form for convenience or for commercial applications. In such cases, the composition must be carefully determined and preferential deposition and even unsuspected impurities are the rule rather than the exception. The composition can be a sensitive function not only of the preparation technique but also of the many deposition parameters and even of the geometry of the system. The development of modern techniques such as Secondary Ion Mass Spectroscopy (SIMS) has been an immense aid in determining the composition, but the important point is the necessity of such a determination.

Once the composition is known, two other non-trivial investigations should be performed. First, it is vital to determine the homogeneity of the material. Macroscopic phase separation over regions greater than 1000 Å is relatively straightforward to identify, but inhomogeneities on smaller scales may be much more subtle. Aggregation of one of the species, such as hydrogen in amorphous silicon-hydrogen alloys, could be important on a scale of the order of only 10 Å or so. Fitting transport data taken on an inhomogeneous film to a homogeneous model is clearly an exercise in futility, yet this may have been done in the majority of papers in the amorphous-silicon field to date.

A related issue is that of possible anisotropy. It seems clear that all thin films have an interface region near the substrate which very likely is significantly different from the bulk. Similarly, an upper interface region must also exist, near a contact material or a gas. If these interface regions are a small fraction of the film thickness, it might be appropriate to ignore them, but this need not be the case. In any event, such regions clearly occupy a greater fraction of the volume of a thin film than in the case of a bulk three-dimensional solid. In addition to these lateral anisotropies in thin films, anisotropy perpendicular to the substrate is often observed. At high deposition powers, columnar growth has recently been observed in amorphous silicon-hydrogen alloys. Once again, it is evident that gross anisotropies in physical parameters can result.

Only after homogeneity and isotropy are established should our attention be turned to the details of the microscopic structure. It is at this point that the amorphous nature of the solid becomes an issue. The appropriate parameter is the range of positional and compositional order. If the material has only one component, there is no compositional disorder. In such cases, it is almost always found that at least the nearest-neighbor coordination of an atom is the same in the corresponding amorphous and crystalline phases. For example, in amorphous silicon, almost every silicon atom is surrounded by four neighboring atoms 2.35 Å away, exactly the same atomic separation as in the crystal. Even the second neighbors are essentially at the same distance as in the crystal, but there is a spread in separations which reflect bond-angle deviations of about ±10° [1]. This spread becomes more and more significant as the distance from the central atom increases, and at distances of the order of the tenth neighbors or so tha atoms are located essentially at random. Thus, pure amorphous silicon can be said to possess short-range order up to about 5 Å. Similar conclusions can be drawn with regard to stoichiometric binary alloys and even some more complex systems. On the other hand, multicomponent alloys, such as the well-studied switching material $Te_{40}As_{35}Si_{15}Ge_7P_3$ [2], do not even possess much nearest-neighbor order because of their compositional disorder. However, even in this case, a property we might call chemical order is present. Because amorphous semiconductors are predominantly covalently bonded, the lowest-energy local configuration of each atom is almost always one in which its chemical valence requirements are fulfilled. Thus, each atom in Columns I-IV in the Periodic Table very likely have 1-4 nearest neighbors, respectively, while each atom in Columns V, VI, and VII generally have coordination numbers of 3, 2, and 1, respectively, in accordance with the 8-N rule of chemical bonding. Any deviation from the optimal coordination ordinarily would cause a large increase in the total energy of the solid, and is thus suppressed if at all possible.

This analysis focuses on the major reason for the similarities in physical properties between corresponding amorphous and crystalline solids, viz. the short-range order in both cases is determined by the chemistry of the component atoms. If large local strains exist during the process of material preparation, then defect centers with non-optimal coordination can form. Since it turns out that these often control the transport properties of amorphous solids (as they do in doped crystalline solids), I shall discuss them in great detail subsequently. We might expect greater defect densities in amorphous solids of larger average coordination because of the concomitantly greater strains in such cases. This issue is discussed in more depth in the paper by Fritzsche [3].

As the average coordination number decreases, another possibility enters, viz. the range of the order can increase beyond that of second or third nearest neighbors. We might call the existence of positional and compositional order in the 10-100 Å regime intermediate range order. There is recent evidence for such intermediate range order in amorphous Se (which has an average coordination number of only 2) [4] and in amorphous Si-F-H alloys [5]. There is, of course, the nebulous question of how long range the order must be before we should properly call the solid crystalline. Clearly, no well-defined distance can be justified purely on geometric grounds, although the word microcrystalline has often been applied in such cases. However, it is possible that certain physical parameters exhibit sharp transitions as the range of the order is increased, and these may be used to make a more definitive dichotomy between the crystalline and amorphous phases. I shall return to this point in the next section.

2. Electronic Structure of Amorphous Solids

It is difficult to calculate the electronic band structure of amorphous solids from first principles because of the absence of crystalline periodicity. However, for those amorphous materials with the same short-range order as a corresponding crystal, the one-electron density of states of the former can be estimated from that of the latter by the introduction of a non-k-conserving perturbation [6]. This perturbation destroys both the validity of k as a good quantum number and the concept of Brillouin zones. In addition, it removes the van Hove singularities from the density of states and, most important of all, it renders Bloch's theorem invalid.

For a semiconductor, the most significant van Hove singularities are those at the upper edge of the valence band and the lower edge of the conduction band, since these define the energy gap. Application of a non-k-conserving perturbation to the periodic potential of a crystalline semiconductor clearly will shift states from the valence and conduction bands into the previously forbidden gap, thus creating what has come to be known as band tails. We might expect that the extent of the band tails in any amorphous semiconductor is a monotonic function of the amount of disorder, with materials like amorphous silicon or amorphous selenium having relatively sharp band tails but with those like amorphous $Te_{40}As_{35}Si_{15}Ge_7P_3$ having extensive band tails. Such a suggestion was made by Cohen et al. [7] over a decade ago.

It should be noted that the destruction of k as a good quantum number in amorphous solids has profound consequences on the optical properties, independent of the extent of the band tails. Since photons with energies less than 10 eV or so have a very small momentum, k is conserved in optical transitions in crystals. This selection rule is responsible for the relatively weak optical absorption even above the energy gap in crystalline silicon and other semiconductors characterized by an indirect edge (i.e. a crystal in which the top of the valence band possesses a different value of k than does the bottom of the conduction band). The selection rule clearly is inappropriate in amorphous solids, because k is not a good quantum number in the absence of periodicity. In polycrystalline solids, each grain retains its periodicity, and k conservation remains applicable even with a 50 Å grain size. The relative spread in k can be estimated from the Uncertainty Principle to be of the order of

$$\frac{\Delta k}{k_{BZ}} \sim \frac{a}{L} ,$$

where k_{BZ} is the maximum crystal momentum in the first Buillouin zone, a is the size of the primitive cell, and L is the grain size. For a 50 Å grain size, the spread in k is less than 10% and is relatively insignificant. However, for smaller grain sizes we might expect deviations from the k selection rule. The presence of such deviations might provide a more rigorous definition of an amorphous solid than does the range of the order discussed in the previous section. This is also a better criterion than one involving the

16

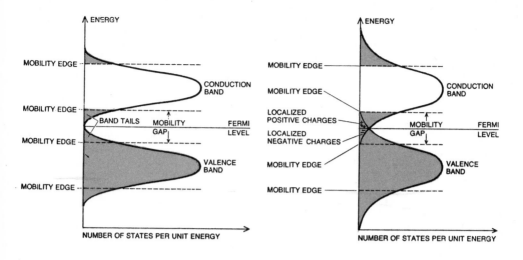

Fig.1 Model of COHEN et al. [7] for the density of states of covalent amorphous semiconductors. The sketch on the left was expected to apply to elemental amorphous semiconductors such as silicon, while that on the right was suggested to be appropriate in multicomponent alloys.

existence of extensive band tails, since band tails can arise from interface states at the grain boundaries of a polycrystalline solid.

Although band tails can have very significant effects on the electronic properties of amorphous semiconductors and the breakdown of the k selection rule can seriously modify the optical properties, the lack of applicability of Bloch's theorem has much more profound consequences for transport. This is because it is Bloch's theorem which is responsible for the extremely high mobilities of free carriers in crystalline semiconductors. Since Bloch states are extended throughout the crystal, free carriers are scattered only by deviations from periodicity. In the absence of long-range order, it is possible that all electronic states are localized in the vicinity of a particular atom. MOTT [8] proposed that a critical density of states exists above which all states exhibit a finite mobility for free carriers even at absolute-zero temperature; this critical density has been called the mobility edge, in analogy to the band edge in crystalline solids. The concept of mobility edges is not valid in one dimension and does not appear to be applicable to two dimensions. However, very little has yet been deduced about the situation in our real three-dimensional world, as is discussed in the paper by WEAIRE [9].

COHEN et al. [7] assumed the existence of band tails and mobility edges, and suggested that for multicomponent amorphous solids the valence and conduction band tails could overlap in the center of the gap, thus yielding a finite density of states at the Fermi energy, $g(\varepsilon_f)$. They also assumed that all the atoms locally satisfied their chemical valence requirements, thus precluding the existence of sharp bumps within the mobility gap, as occurs in doped crystalline semiconductors. Their model, sketched in Fig. 1, provides a useful structure for the analysis of transport and optical data. However, it is useful to ask how the assumptions of the model might be tested by such experiments.

Extensive band tails should be evident in ordinary optical-absorption experiments. In crystalline semiconductors, the material is essentially transparent to light of frequency below that of the energy gap, but the optical-absorption coefficient rapidly increases with photon frequency above the gap. It might be expected that if extensive band tails exist, the optical-absorption coefficient should begin to increase with photon frequencies well below the mobility gap, and this increase should be much less sharp than in the case of crystals with direct edges.

The most straightforward manifestation of a sharp mobility edge would be an electrical conductivity which varies as

$$\sigma = N_c \; e \; \mu_e \; \exp[-(E_c'-\varepsilon_f)/kT] + N_v \; e \; \mu_h \; \exp[-(\varepsilon_f-E_v')/kT], \qquad (1)$$

where N_c and N_v are the effective densities of states in the conduction and valence bands, μ_e and μ_h are the electron and hole mobilities beyond the mobility edges, and E_c' and E_v' are the positions of the conduction and valence band mobility edges, respectively. Eq. (1) would yield a log σ vs. T^{-1} plot that is either a single straight line or two intersecting straight lines. On the other hand, a diffuse mobility edge should yield a concave upward curve on a plot of log σ vs. T^{-1}, reflecting the fact that at lower temperatures the conductivity is dominated by relatively low-mobility carriers much closer to the Fermi energy.

If the suggestion of COHEN et al. [7] that a finite $g(\varepsilon_f)$ exists is correct, the Fermi energy would then appear to be pinned, in the sense that conductivity would be quite insensitive to substitutional doping or to the injection of excess carriers. In addition, the redistribution of electrons which should leave relatively high-energy valence-band states above ε_f (see Fig.1) should create relatively large densities of unpaired spins; these would be expected to be observable in EPR experiments and as a Curie term in the magnetic susceptibility.

MOTT [10] pointed out that for a sufficiently large $g(\varepsilon_f)$, phonon-assisted hopping among localized states near ε_f might well predominate band-like conduction beyond the mobility edges at sufficiently low temperatures. He further suggested that at very low temperatures, when very few energetic phonons are present, such hopping conduction would take place preferentially to farther rather than nearest-neighboring localized states, in order to reduce the energy that must be obtained from the phonons. MOTT showed that for this mechanism of variable-range hopping conduction, the conductivity varies as,

$$\sigma = \sigma_0 \; \exp[-(T_0/T)^{1/4}], \qquad (2)$$

where σ_0 and T_0 are constants. When (2) is obeyed, a log σ vs. T^{-1} plot is concave upward, but a log σ vs. $T^{-1/4}$ plot should be linear.

The possibility of variable-range hopping actually somewhat muddles the test for the existence of sharp mobility edges. A linear plot of log σ vs. T^{-1} could indicate either band-like conduction beyond a sharp mobility edge or phonon-assisted hopping between nearest-neighboring localized sites. Alternatively, a concave upward log σ vs. T^{-1} plot could reflect either the absence of a sharp mobility edge or the predominance of variable-range hopping conduction.

In principle, field-effect measurements provide an excellent probe of the density of states near the Fermi energy. This technique involves the injection into the semiconductor of electrons or holes, which should then cause ε_f to shift upwards or downwards, repectively. The extent of this shift can be determined from the corresponding change in conductivity, and it can then be related directly to g(E). In principle, the field-effect technique can also be used to determine if g(E) is smooth or exhibits structure, the latter indicating a significant lack of local valence satisfaction.

To summarize, a priori predictions of COHEN et al. [7] for amorphous Si, Ge, and Se suggest that these semiconductors should exhibit a large field effect, the possibility of substitutional doping, a small density of unpaired spins, no variable-range hopping, and a sharp absorption edge. In contrast, a multicomponent glass such as $Te_{40}As_{35}Si_{15}Ge_7P_3$ should be characterized by a small field effect, no substitutional doping, a large unpaired-spin density, variable-range hopping, and a diffuse absorption edge.

When all of these experiments were actually carried out, many surprises emerged. For example, pure amorphous silicon deposited on a room-temperature substrate exhibits none of the predicted properties: there is no observable field effect, no doping, a large unpaired-spin density, variable-range hopping conduction, and a diffuse absorption edge. Many of these properties, however, are not intrinsic and anneal away at temperatures below the crystallization point. On the other hand, when amorphous silicon is hydrogenated either during deposition or afterwards, the expected behavior is observed in each case. We can conclude that a purely tetrahedrally coordinated amorphous network necessarily is sufficiently strained that large defect densities exist. However, these can be reduced by annealing. Furthermore, a modification of the network by the introduction of a monovalent component such as hydrogen serves to reduce the strains sufficiently to remove almost all of the defects.

The properties of multicomponent chalcogenide glasses are much more anomalous. In almost all of these semiconductors, the Fermi energy appears to be very strongly pinned, as evidenced by the absence of either any doping effects or an observable field effect under ordinary conditions. On the other hand, no variable-range hopping or any unpaired-spin density has been observed except under the most non-equilibrium conditions, and the absorption edge is not significantly more diffuse in the multicomponent glasses than it is in amorphous Se or Te. Relatively large field effects have been observed in multicomponent glasses containing Te and As, but these are transient and decay with time. Thus, there is evidence both for and against a large density of states at the Fermi energy, and the experimental results appear to be contradictory.

3. Classification of Amorphous Solids

The key to understanding the electronic properties of covalent amorphous semiconductors is the recognition of two major facts: (1) the classification should not be one which emphasizes the extent of the disorder present, but rather the chemical nature of the constituent atoms and the type of bonding; and (2) the transport properties are not ordinarily controlled by band-tail states, but rather by the localized states originating from defect centers with non-optimal local coordination. Although band tails must exist in disordered system, they need not be extensive, and there are theoretical reasons for believing it is unlikely that the valence and conduction band tails ever overlap sufficiently to yield a very larg $g(\varepsilon_f)$ [11]. Thus, the observation of variable-range hopping conduction at ordinary temperatures probably reflects large defect densities rather than an intrinsic property of the semiconductor, in accordance with the results on amorphous silicon films.

Let us limit our discussion to atoms with outer s and p electrons; the maximum possible number of covalent bonds per atom is thus four. If the solid consists of atoms with an average of four outer electrons, e.g. Si, Ge, GaAs, CdS, then tetrahedral (sp³) bonding is optimal. However, as discussed previously, internal strains upon film deposition often results in large defect densities. These can be reduced by depositing the film on a high-temperature substrate or by annealing the film after deposition, but such procedures are limited by the crystallization temperature of the material. Because the defects are not intrinsic to the material but depend on the deposition conditions, it is extremely difficult to predict the physical properties a priori.

When the average number of outer electrons differs from four, pure tetrahedral bonding is no longer optimal. A simple example is the amorphous silicon-hydrogen system. Hydrogen can bond with only one other atom (ignoring the possibility of weak hydrogen bonding), and thus its introduction into silicon lowers the average coordination considerably. This serves to reduce the strains in the films and concomitantly lower the defect density. This is borne out by extensive observations. Similarly, introduction of one of the halogen (Column VII) elements also lowers the average coordination by the same amount as a comparable concentration of hydrogen. The bonding in this case ranges from a largely ionic bond ($Si^+ - F^-$) in the case of fluorine to a predominantly covalent bond in the case of iodine; the latter alloys may even prefer the formation of three-fold-coordinated silicon and doubly coordinated iodine. Nevertheless, the defect density should be reduced considerably unless atomic-size constraints become important. Alloying of silicon or germanium with chalcogen (Column VI) elements, which are divalent, reduces the average coordination more slowly than hydrogen or halogens. In this case, either the Column IV atoms retain their tetrahedral environment and the chalcogens retain their p² bonding, or three-fold-coordinated Column IV atoms can be paired with three-fold-coordinated chalcogens, depending on the particular system as well as the preparation conditions. But in any case, the lower average coordination should greatly reduce the defect density.

Pnictide (Column V) amorphous semiconductors such as arsenic or phosphorus optimally bond in a p³ configuration, with three nearest neighbors. This configuration introduces much less strain into the network than does tetrahedral bonding, and we would expect a much lower defect density. This is in agreement with the available experimental data. Similarly, chalcogenide semiconductors should exhibit predominantly two-fold coordination, and consequently minimal strain upon formation; we might thus expect very small defect densities. Unfortunately, this is not consistent with the experimental data. As discussed in the previous section, the Fermi energy in essentially all of the chalcogenides appears to be strongly pinned, characteristic of very large densities of defects.

Clearly, the defect density is not only a function of the strains in the material during preparation. Defect centers can also result from thermodynamic considerations, provided the formation energy of the defect, ΔE, is small. Glasses are amorphous semiconductors which can be quenched from the liquid into a disordered structure; they become solid at a well-defined glass transition temperature, T_g, below which the viscosity increases by many orders of magnitude. Such materials must thermodynamically exhibit a minimum defect density,

$$N_D \geq N_o \exp (-\Delta E/kT_g), \qquad (3)$$

where N_o is the density of the material. For $\Delta E = 0.5eV$ and $T_g = 600$ K, (3) shows that the defect density must be at least of the order of $10^{19} cm^{-3}$. As I shall discuss in the next section, low-energy defects always exist in chalcogenide alloys, and because of their unusual nature they control the transport properties of the materials.

5. Defects in Amorphous Semiconductors

5.1 Experimental Evidence for Defects

Although many of the defects common in crystalline solids, e.g. vacancies, interstitials, dislocations, and grain boundaries, are not likely to be found in the absence of periodicity, chemical defects, such as undercoordinated or overcoordinated atoms and wrong bonds, are quite possible. Before analyzing the nature of such defects, we might well ask whether there is any compelling experimental evidence for their existence.

Perhaps the most convincing set of experiments which cannot easily be understood without the presence of well-defined defects are the photoluminescence results on chalcogenides. STREET [12] analyzed the excitation spectrum for photoluminescence in several chalcogenide glasses. He noted that the spectrum falls off on the low-energy side because of the decrease in absorption. However, this cannot explain a similar fall-off observed on the high-energy side. STREET showed that the latter was not due to surface recombination, but, in fact, was a bulk phenomenon, the decrease in luminescence efficiency being inversely proportional to the absorption constant,

$$n(E) \propto \frac{1}{\alpha(E)} . \qquad (4)$$

Such a relationship is most easily explained if definite luminescence centers exist, such that the luminescence occurs only if the incident photons are absorbed in their vicinity. Then, we can write the quantum efficiency as

$$\eta(E) = \frac{\alpha_c(E)}{\alpha(E)} , \qquad (5)$$

where $\alpha_c(E)$ is the absorption coefficient near the luminescence centers. Since we might expect $\alpha_c(E)$ to be essentially independent of energy, (4) follows immediately.

Recently, GEE and KASTNER [13] performed a set of extremely convincing luminescence experiments on crystalline and amorphous SiO_2. Amorphous SiO_2 has an absorption edge in the vicinity of 10 eV. However, the photoluminescence excitation spectrum peaks near 7.6 eV, clearly separated from the continuum, as is evident from Fig. 2. Also shown in Fig. 2 is the photoluminescence, excitation, and absorption spectra for both amorphous As_2S_3 and amorphous SiO_2, normalized to the energy at which the absorption coefficient is $\alpha = 10^4 cm^{-1}$. The similarities in data for two such different glasses is strong evidence in favor of a similar mechanism for the luminescence.

21

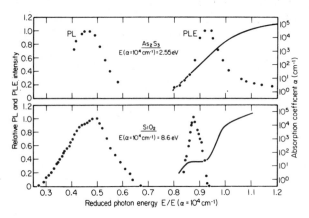

GEE and KASTNER [14] also found that the photoluminescence and excitation
spectra for both neutron-irradiated crystalline and amorphous SiO_2 are essentially the same. On the other hand, whereas the temperature dependence of
the luminescence intensities of unirradiated a-As_sS_3 and a-SiO_2 are identical
when the temperature is scaled to the glass transition temperature, irradiated
a-SiO_2 has a luminescence 20 times stronger. Irradiated samples of crystalline SiO_2 have a strong absorption peak at 7.6 eV, which can result from
nothing but a defect center. After irradiation, the excitation spectrum
still peaks near 7.6 eV and the luminescence still peaks in the 4.5-5.0 eV
range, but the luminescence is greatly enhanced. Obviously, irradiation
simply increases the density of defects already present in the unirradiated
samples, and these defects are responsible for the luminescence. In addition, the similarities in the scaled temperature quenching of the photoluminescence and the scaled energy dependence of both luminescence and excitation spectra for several different glasses suggest a common mechanism for
the entire process. We are thus led to conclude that well-defined defects
similar to those in the corresponding crystals are responsible for photoluminescence in chalcogenide glasses.

There are many other types of experiments which also indicate the presence of defect centers. For example, the drift mobility for holes in amorphous As_2Se_3 is trap-limited with an activation energy of 0.60 eV. When
concentrations of up to $10^{19}cm^{-3}$ of Tl are incorporated in the material, the
density of hole traps sharply increases, but the activation energy remains
at 0.60 eV [15]. Thus, Tl incorporation simply increases the concentration
of those defect centers which trap holes. As will be discussed shortly, the
hole traps in amorphous chalcogenides are expected to be negatively charged,
singly coordinated chalcogens. In pure As_2Se_3, such centers must be balanced
by positively charged, three-fold-coordinated chalcogens. However, when Tl
is introduced, the energy of the least-bound electron on the Tl is sufficiently large that electrons leave the Tl atoms and, in pairs, convert positively to negatively charged chalcogens (with an additional bond-breaking
in the process); this creates additional hole traps while maintaing the
same trap energy, in accordance with the observations of PFISTER et al. [15].

There has been a great deal of work in recent years concerned with the phenomenon of threshold switching [16] in chalcogenide-glass films. When a critical field is applied across these materials, there is a sharp increase in free-carrier concentration from values of the order of $10^{11} cm^{-3}$ to approximately $10^{19} cm^{-3}$. The detailed results [17] indicate the predominance of hole conduction in the high-resistance state but electron conduction in the low-resistance state, propagation of a plasma from anode to cathode prior to switching, and persistance of the cathode field after removal of voltage from a sample in the low-resistance state. All of these phenomena can be explained in a straightforward way by the existence of pairs of well-defined charged defects [2]. In addition, electroluminescence from the low-resistance state has been observed and is similar in appearance to the photoluminescence in a wide class of chalcogenide glasses [18].

Experimental evidence for the existence of defects in amorphous semiconductors is not restricted to chalcogenides. There is also an abundant body of data on tetrahedrally bonded solids. One of the simplest examples is the apparent substitutional doping. If phosphorus, arsenic, and boron do enter amorphous silicon-based alloys in tetrahedral sites, such centers then have the wrong chemical coordination and represent defects. In an approach such as that of COHEN et al. [7], in which each atom locally satisfies its valence requirements, this type of doping would not be possible. But even in undoped samples, the existence of defects can be inferred unambiguously. Fig. 3 shows the results of FRITZSCHE et al. [19], in which it is clear that the spin density of irradiated amorphous silicon-hydrogen alloys decreases with increasing substrate temperature until about 350°C. However, as samples are annealed above 350°C, hydrogen effuses and the spin density sharply increases. When hydrogen effuses, defect centers must be created.

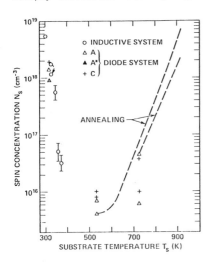

Fig. 3 Spin concentration as a function of substrate temperature and annealing for amorphous silicon-hydrogen alloys [19].

The spin density was observed by EPR measurements, which indicated a common center for all spins. Clearly, samples prepared at low substrate temperatures must contain some of the same defect centers as those created when hydrogen effuses. We can conclude that well-defined defects are also present in tetrahedrally coordinated amorphous solids.

5.2 Defects in Chalcogenides

KASTNER et al. [20] showed that a particularly low-energy defect exists in chalcogenide alloys, a valence alternation pair (VAP), consisting of a positively charged three-fold-coordinated chalcogen and a negatively charged singly coordinated chalcogen. In their notation, this pair is represented $C_3^+ - C_1^-$. The creation energy of a VAP is U, the additional Coulomb repulsion arising from the presence of the extra electron on the C_1^- site. U represents the difference between the ionization potential of the chalcogen atom and its electron affinity, screened by the dielectric response of the material. Crude estimates for chalcogen atoms suggest that U is of the order of 0.5-1.0 eV so that we might expect $10^{16}-10^{20} cm^{-3}$ VAPs in chalcogenide glasses under ordinary conditions.

The reason for the low creation energy of a VAP is that the total number of bonds in a $C_3^+ - C_1^-$ pair is four, exactly the same as in a pair of chalcogen atoms in their ground states (C_2^0). Furthermore, both C_3^+ and C_1^- centers have low energy: a positively charged chalcogen is isoelectronic with a pnictide, and therefore it is optimally three-fold coordinated; a negatively charged chalcogen is isoelectronic with a halogen, and thus is optimally singly coordinated.

As discussed previously, the existence of these low-energy defects guarantees their presence in large concentrations under all preparation conditions. But the VAPs also have a very unusual property which resolves the puzzle of why the Fermi energy is so strongly pinned in chalcogenides while at the same time there is no measurable unpaired-spin density and variable-range hopping conduction is not observed in any temperature range. This will be discussed in detail in the next section, but its origin is the fact that the two defect centers, C_1 and C_3, have coordination numbers separated by two. This provides the possibility of converting a C_3 to a C_1 center simply by the breaking of a bond (since breaking of a two-electron bond lowers the total coordination of the atoms in a network by two). It is this possibility, together with the fact that a positively charged C_3 center and a negatively charged C_1 center have lower total energy than two neutral centers with either the C_1 or C_3 configuration, that leads to the Fermi-energy pinning. The presence of VAPs also explain a great deal of other unusual characteristics of chalcogenide glasses including the complete absence of spins (neither a C_3^+ nor a C_1^- center contains any unpaired spins), the lack of observation of variable-range hopping (two electrons or holes must hop simultaneously to interconvert C_3^+ and C_1^- centers, and the resulting electrostatic relaxations cost too much energy), the presence and the details of the photoconductivity, photoluminescence, and photostructural effects [20]. As will be discussed shortly, there are different types of defect centers in other amorphous semiconductors that can pin the Fermi energy and act as traps for free carriers. However, what is virtually unique about chalcogenides is the low creation energy of such defects, and this follows from the existence of a lone pair in the outermost filled sub-shell of chalcogen atoms [21,22].

5.3 Defects in Amorphous Silicon and Related Alloys

Because four is the maximum possible covalent coordination using s and p orbitals only, valence alternation is impossible in tetrahedrally bonded solids. However, as previously discussed, the three-dimensional structural rigidity of tetrahedral bonding together with the usual techniques used to deposit such amorphous films ensure the existence of large defect densities. The lowest-energy neutral defects are the two-fold and the three-fold coordinated atoms, T_2^0 and T_3^0, respectively. Crude estimates [23,24] suggest that the T_2^0 center has lower energy than the T_3^0, primarily because of the fact that s-p promotion is not necessary for T_2^0 center. However, the creation energy for both defects is quite large, and we can conclude that they form <u>only</u> because of strains during the preparation of the sample.

Although the high formation energy of defect centers in these solids suggests that relatively defect-free films can be formed in principle, large defect densities appear to exist in all purely tetrahedrally bonded amorphous solids. Thus, it is significant to ask if either T_2 or T_3 defects pin the Fermi energy as do VAPs in chalcogenides. It is clear [24] that T_2 centers do <u>not</u>, but we shall show in the next section that T_3 centers may well do so in a-Si. If this is the case, T_3^+- T_3^- pairs form in the material. Either a T_3^+ center (the lowest-energy configuration for an atom with three outer s-p electrons) or a T_3^- center (the lowest-energy configuration for an atom with five outer s-p electrons) has all of its spins paired in either bonding orbitals or lone pairs. Thus, they are spinless defects, like C_3^+ and C_1^- in chalcogenides or T_2^0 in a-Si itself.

There is a great of experimental evidence for the existence of large concentrations of spinless defects in amorphous-silicon films. The observed spin density is almost always several orders of magnitude lower than the density of localized states in the gap [25,26]. When atomic hydrogen is introduced into pure silicon films, approximately 100 times as much hydrogen enters than the unpaired-spin density suggests [27]. These spinless defects could be strained bonds, T_3^+ - T_3^- pairs, or T_2^0 centers. Of these, only the T_2^0 center could easily pick up atomic hydrogen without the necessity of significant atomic relaxation. Futhermore , the presence of large concentrations of T_3^+ - T_3^- pairs would tend to suppress variable-range hopping conduction, as VAPs do in chalcogenides. However, linear plots of log σ vs. $1/T^{1/4}$ are routinely observed in a-Si films [28]. Thus, the T_2^0 center may be the most likely mechanism for the spinless defects.

Much more information is available on amorphous silicon-hydrogen alloys, particularly those films produced by the glow-discharge decomposition of SiH_4 gas [26]. The resulting alloys contain 10-35% hydrogen, depending primarily on the substrate temperature. At low substrate temperatures, larger concentrations of hydrogen are generally present. KNIGHTS [29] has shown that this spin density is correlated with a columnar growth, presumably due to the presence of $(SiH_2)_n$ chain segments (polysilane). At higher substrate temperatures, up to about 300°C, the hydrogen that is incorporated is generally bound to different silicon atoms, and the spin density decreases. Above this point, the spin density begins to increase, presumably reflecting an effusion of hydrogen during deposition. This situation, high spin densities and low Si-H_2 concentrations, can also be attained at low substrate temperatures by depositing films on biased surfaces [29]. When amorphous silicon-hydrogen alloys are annealed, hydrogen gas is given off, first from Si-H_2 sites, later from Si-H sites [19]. The unpaired-spin density increases with this effusion, although the order of 100 hydrogen atoms are given off per unpaired spin produced [26].

All of these results can be analyzed using a defect model. The silicon-hydrogen bond energy is 3.4 eV, significantly larger than 2.4 eV silicon-silicon bond energy. Since hydrogen can be only singly coordinated (not including the possibility of either hydrogen bonding or the formation of three-center bonds, which occurs when boron is present), it tends to relieve the mechanical strains that accompany a purely tetrahedral amorphous network. This should sharply reduce the density of defect centers that form upon deposition, thereby removing localized states from the gap. Nevertheless, there are still 10^{18}-10^{19} cm^{-3} localized states remaining in the gap, far more than the observed unpaired-spin density of 10^{16}-10^{17} cm^{-3} [26].

Once again, there are undoubtedly spinless defects present. An obvious possibility is again the T_2^0 center. This must be created when hydrogen effuses from an Si-H$_2$ site, and its formation is consistent with the previously mentioned observation that about 100 times as many hydrogen atoms effuse for every unpaired spin created during such experiments. However, other possibilities should be considered. When there are regions of polysilane present, effusion of a molecule of H$_2$ from a single silicon site immediately yields a T_2^0 center. However, transfer of an H atom originally bound to a neighboring silicon site is equivalent to the reaction, $T_2^0 \rightarrow 2T_3^0$. Furthermore, as discussed previously, charge transfer can then take place, yielding the overall reaction, $T_2^0 \rightarrow T_3^+ + T_3^-$ [30]. Clearly, the situation can be quite complex.

Unfortunately, tight-binding calculations [23] compound the complication still further by suggesting that another relatively low-energy charged defect exists, a T_2^+ - T_3^- pair. This is an interesting defect pair in that it contains an unpaired spin on the T_2^+ center. In addition, the charged centers would act as efficient electron and hole traps, which would limit the transport of injected or photogenerated carriers. The presence of such centers is also consistent with a great deal of other experimental data [11].

6. Effective Correlation Energy of Defect Centers

Conventional band theory neglects the possibility that two electrons can correlate their motion to minimize their mutual repulsion. Consequently, the total energy is ordinarily overestimated by a term proportional to the repulsion between two electrons simultaneously located on the same atomic site, a term usually called U. In extended states, the screening of the Coulomb interaction by the mobile carriers reduces the effective value of U sufficiently that no serious problems result from the neglect of correlations. However, screening is not nearly as effective in localized states, and more care is necessary.

Significant consequences of the existence of electronic correlations in localized states occur in as common a problem as that of phosphorus-doped silicon. The so-called donor state of a phosphorus atom in tetrahedral coordination is essentially the 1s orbital of a hydrogen atom (formed by the extra electron of the phosphorus atom moving in the conduction band of the silicon atoms and attracted to the P$^+$ ion core). Since the 1s orbital is doubly degenerate, two electrons can simultaneously occupy each donor level. If the two electronic states were located at exactly the same energy, the donor levels in Si:P would be half filled and conduction could occur without an activation energy. However, because of the extra Coulomb-repulsion, the energy levels of the unoccupied 1s orbitals associated with a neutral phosphorus atom (i.e. with the other 1s orbital occupied) is located an energy U above those of the occupied orbitals. Consequently, the phosphorus atoms in crystalline silicon at very low temperatures are all neutral. Note that U can be looked at as the energy necessary to create a P$^+$ - P$^-$ charged pair.

In general, all defect centers have an associated value of U, which is the energy necessary to create a charged pair,

$$2D^0 \rightarrow D^+ + D^-. \tag{6}$$

Of course, atomic relaxations can take place around both the D^+ and the D^- centers, lowering the <u>effective</u> value of U somewhat; the actual energy increase due to reaction (6) is thus called U_{eff}.

We might ask how the Fermi energy varies with injection of excess electrons or holes, an important function in the analysis of field-effect experiments. If there is only one type of defect center in a particular solid, then the Fermi energy at equilibrium is located an energy $U_{eff}/2$ above the levels associated with singly occupied defects. However, excess electrons form D^- centers, thus increasing the Fermi energy by $U_{eff}/2$, while excess holes form D^+ centers, reducing the Fermi energy by $U_{eff}/2$. The situation is sketched in Fig. 4.

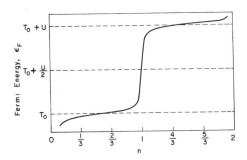

Fig. 4 Fermi energy as a function of electron concentration per defect center for a positive effective correlation energy, U.

Consider the dangling bond in amorphous silicon. When a Si atom has only three nearest neighbors, the eight-fold degenerate sp^3 levels are split such that three filled bonding orbitals have low energy, three empty antibonding orbitals have high energy, and two nonbonding levels lie in between. If the center is locally neutral, one of these is filled, the other empty. The nonbonding electron represents an unpaired spin which should contribute to the EPR signal. Because of the Coulomb repulsion, the empty nonbonding level has an energy U above that of filled level in the absence of relaxations. However, relaxations do exist, and U is not a good guide to the value of U_{eff}.

In its usual tetrahedral configuration, Si forms sp^3 bonds, with a bond strength, E_b. If one bond is dangling, the energy of the Si atom relative to one with four nonbonding sp^3 electrons is then $-3E_b$. If one electron is removed from a three-fold coordinated Si atom, it is the nonbonding electron, and the energy of the Si^+ center remains $-3E_b$. If this electron is now placed on a second three-fold coordinated Si atom, thus forming a Si^- center, a dehybridization can occur with a reduction of the bond angle from the tetrahedral 109.5° to that optimal for predominantly p bonding (90°-100°). The three p bonds are weaker than the sp^3 bonds, but this is more than regained by recapture of the s-p promotion energy, P. If B represents the excess energy of an sp^3 bond relative to a p bond, the energy of a Si^- center is

$$-3E_b - \frac{3}{4}(P - 4B) + U, \tag{7}$$

27

where we have taken into account the correlation energy U necessary to form the Si⁻ ion. Thus, the energy necessary for the reaction

$$2Si \rightarrow Si^+ + Si^-$$ (8)

is

$$U_{eff} = U - \frac{3}{4}(P - 4B).$$ (9)

The quantity (P - 4B) must be positive, since otherwise Column V atoms would exhibit predominantly sp³ rather than p³ bonding, at variance with observations. In fact, (P - 4B) can be estimated to be quite large, perhaps 1.5 eV or more [23]. If so, it is likely that $U_{eff} < 0$ for the dangling bond in Si, as suggested in the last section.

There are many important physical consequences of a negative effective correlation energy. Neither Si⁺ nor Si⁻ contains unpaired spins, since both centers are optimally bonded. Furthermore, the unpaired-spin density remains vanishingly small even if excess electrons and holes are injected into the material. This is because a pair of excess electrons preferentially fall on the same Si⁺ center, converting it to an Si⁻ center. (If U_{eff} is negative, then an Si⁺ - Si⁻ pair has lower energy than two neutral dangling bonds.) Conversely, injection of a pair of holes simply converts an Si⁻ to an Si⁺ center. Since neutral centers are unstable, unpaired spins do not form at equilibrium.

It is also straightforward to see that a negative U_{eff} pins the Fermi energy. Every pair of electrons injected into the material enters with exactly the same energy, since they all induce the reaction,

$$Si^+ + 2e^- \rightarrow Si^-.$$ (10)

Since the Fermi energy measures the average energy needed to add a small number of excess electrons to the solid, it does not vary with electronic density if $U_{eff} < 0$. The results of a detailed calculation are shown in Fig. 5.

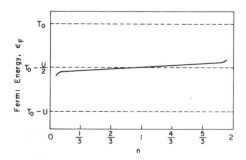

Fig. 5. Fermi energy as a function of electron concentration per defect center for the case of a negative effective correlation energy, U.

If the dangling bond in amorphous Si has a negative effective correlation energy, then the origin of the observed EPR signal in many samples of the material must be identified. One possibility is the T_2^+ center discussed in the last section. Another explanation will be analyzed in the next section.

In chalcogenides, no EPR signal is ordinarily observed. It is straight-forward to understand the origin of the negative U_{eff} in these materials [20]. Two neutral defects are possible with chalcogen atoms, an overcoordin-ated atom (C_3^0) and a dangling bond (C_1^0). Only p bonding is involved in either case. The creation energy for an overcoordinated center is the bond-ing-antibonding energy loss, Δ. If an electron is removed from a C_3^0 center, it is the antibonding electron, and the energy is reduced by $E + \Delta$; if that electron is placed on a dangling bond, it enters a nonbonding orbital, and the energy is increased by U. Thus, the energy for the reaction,

$$C_3^0 + C_1^0 \to C_3^+ + C_1^-, \tag{11}$$

Is $U - (E + \Delta)$, almost certainly negative. In fact, the creation energy of a C_3^+ - C_1^- pair is only U, a fact responsible for the large defect density in chalcogenides, as discussed previously. But this is not sufficient to understand the Fermi energy pinning since a C_1 center and a C_3 center are not the same. The unusual properties of chalcogenide glasses result from the possibility of effectively interconverting C_1 and C_3 centers by break-ing or forming a bond. For example, if one of the extra bonds on a C_3 center breaks in amorphous Se, the C_3 center returns to the ground (C_2) state, but the neighboring C_2 site becomes a dangling bond (C_1). The net effect is

$$C_3 \to C_1. \tag{12}$$

Formation of an extra bond around a dangling bond induces the reverse of (12). It is this possibility which pins the Fermi energy; injection of two electrons into the material simply induces

$$C_3^+ + 2e^- \to C_1^- \tag{13}$$

after the necessary bond breaking. Once again, no spins are introduced by the injection.

In chalcogenide alloys, such as As_2Se_3, the lowest-energy defect remains a VAP, most likely a C_3^+- P_2^- pair (where P stands for a Column V or pnic-tide atom). Injection of a pair of electrons together with a bond breaking induces the net reaction,

$$C_3^+ + 2e^- \to P_2^-. \tag{14}$$

Thus, it is not surprising that the major properties of As_2Se_3 and Se are quite similar.

7. Transient Effects

We might ask if the existence of defects with a negative correlation energy has any effect on non-equilibrium experiments such as transient photocon-ductivity, drift mobility, or field effect. In all of these cases, we might expect rapid trapping of the excess electrons and/or holes by the charged centers and a relatively rapid return to equilibrium. However, this is not necessarily the case. For example, a detailed analysis of the process (14) shows that it takes place in the three steps: (1) trapping of an electron by the positively charged center,

$$C_3^+ + e^- \to C_3^0 ; \tag{15}$$

(2) relaxation involving a bond breaking,

$$C_3^0 \rightarrow P_2^0;$$ (16)

and (3) trapping of a second electron by the neutral center,

$$P_2^0 + e^- \rightarrow P_2^-.$$ (17)

When excess free electrons are injected into, e.g. As_2Te_3 via a field-effect experiment, all three steps (15)-(17) are necessary in order to achieve re-equilibration. Only if the time scale of the observation is long compared to that of the longest of the three processes will the pinning of the Fermi energy manifest itself. The most likely bottleneck is (16), since it is probable that both C_3^0 and P_2^0 are local minima in the structure of the solid; thus, a potential barrier should exist between the two defect centers, and an appropriate Boltzmann factor will enter the equilibration kinetics [30]. For very short times, C_3^+ centers trap electrons and P_2^- centers trap holes, but the two neutral centers do not interconvert. Thus, the apparent density of states is that of a doped, compensated semiconductor. The Fermi energy is not pinned on this time scale, and a transient field effect should be observable. However, this field effect should decay with time, the time constant yielding the height of the potential barrier. Such transient effects have been observed in Te-As glasses and have been quantitatively analyzed [30]. Clearly, the same type of transients should manifest themselves in photoconductivity experiments, except for the added requirement of charge neutrality. We shall return to this point shortly.

The relaxation time for (16) should be given by

$$\tau \simeq \tau_{ph} \exp[\Delta E/kT],$$ (18)

where ΔE is the height of the barrier and τ_{ph} is a phonon time, of the order of 10^{-13} sec. Thus, at room temperature, with a ΔE of 0.5 eV, τ is about 50 μsec. However, for a ΔE of 1.0 eV, τ is of the order of several hours. Clearly, these transient effects can persist for very long times. This is the case for glasses in the Te-As system [30].

One potential difficulty looming over the application of amorphous silicon-hydrogen alloys as semiconductor devices is the existence of a metastable state, first observed by STAEBLER and WRONSKI [31]. Whereas as-deposited films exhibited a room-temperature conductivity of the order of $10^{-6}\Omega^{-1}cm^{-1}$ and an activation energy of 0.57 eV, after the film was exposed to light, the conductivity decreased to about $10^{-10}\Omega^{-1}cm^{-1}$ and the activation energy increased to 0.87 eV. The original state can be restored by annealing the film for two hours at 150°C. The relaxation time for recovering the as-deposited state has an activation energy of about 1.5 eV, so that the room-temperature relaxation time is over 25,000 years.

The existence of charged defect centers provides a natural explanation of such effects [30], since electrons can be trapped at the positive centers forming metastable state. In fact, the crude estimates discussed in the last section suggest that these charged centers could well be T_3^+ - T_3^- pairs. Although such pairs may have a negative U_{eff}, the presence of a very high potential barrier between the sp^3 and p^3 configurations of the neutral dangling bond would result in metastable unpaired spins. Although the validity of such an explanation remains to be verified experimentally, in any event, the origin of the STAEBLER-WRONSKI effect must be identified before the electronic structure of amorphous silicon can be considered to be understood.

8. Conclusions

At the present state of our knowledge of covalent amorphous semiconductors, the following ten points appear to be able to explain the available data:

(1) Well-defined defects exist in covalent amorphous semiconductors.

(2) These ordinarily control the equilibrium transport properties of the material via the position of the Fermi level.

(3) They can also be expected to control the trapping and recombination kinetics.

(4) Several different types of defect centers can be simultaneously present in any particular solid.

(5) The defect centers can be either charged or neutral.

(6) The effective correlation energy can be positive or negative.

(7) A positive effective correlation energy allows the Fermi energy to be moved through the gap by the magnitude of U_{eff} with electron or hole injection.

(8) A negative effective correlation energy pins the Fermi energy to the extent of the density of that particular defect.

(9) If U_{eff} is negative, a potential barrier between the two possible neutral defects can lead to important transient effects.

(10) These transient effects can persist for a time sufficiently long that the neutral defects can be considered to be metastable even at room temperature.

Acknowledgments

I should like to thank Robert C. Frye, Marc Kastner, Stanford R. Ovshinsky, and Marvin Silver for useful conversations. The original research discussed was supported by grants from the U. S. Army Research Office and the U. S. National Science Foundation Materials Research Laboratory Program.

References

1. S. C. Moss and J. F. Graczyk, in Proc. Tenth Intern. Conf. on Phys. of Semicond. (U.S.A.E.C., Washington, 1979), p. 658.
2. D. Adler, M. S. Shur, M. Silver, and S. R. Ovshinsky, J. Appl. Phys. 51, 3289 (1980).
3. H. Fritzsche, this volume.
4. T. Takahashi, K. Ohno, and Y. Harada, Phys. Rev. B 21, 3399 (1980).
5. R. Tsu, M. Izu, S. R. Ovshinsky, and F. H. Pollak, Solid State Commun., in press.
6. D. Adler, CRC Crit. Rev. Solid State Sciences 2, 317 (1971).

7. M. H. Cohen, H. Fritzsche, and S. R. Ovshinsky, Phys. Rev. Lett. $\underline{22}$, 1065 (1969).
8. N. F. Mott, Adv. Phys. $\underline{16}$, 49 (1967).
9. D. Weaire, this volume.
10. N. F. Mott, Phil. Mag. $\underline{19}$, 835 (1969).
11. D. Adler, Solar Cells, in press.
12. R. A. Street, Adv. Phys. $\underline{25}$, 397 (1976).
13. C. M. Gee and M. Kastner, Phys. Rev. Lett. $\underline{42}$, 1765 (1979).
14. C. M. Gee and M. Kastner, J. Non-Crystall. Solids $\underline{35/36}$, 807 (1980).
15. G. Pfister, K. S. Liang, M. Morgan, P. C. Taylor, E. J. Friebele, and S. G. Bishop, Phys. Rev. Lett. $\underline{41}$, 1318 (1978).
16. S. R. Ovshinsky, Phys. Rev. Lett. $\underline{21}$, 1450 (1968).
17. D. Adler, H. K. Henisch, and N. F. Mott, Rev. Mod. Phys. $\underline{50}$, 209 (1978).
18. P. J. Walsh, S. Ishioka, and D. Adler, Appl. Phys. Lett. $\underline{33}$, 593 (1978).
19. H. Fritzsche, C. C. Tsai, and P. Persans, Solid State Tech. $\underline{21}$, 55 (1978).
20. M. Kastner, D. Adler, and H. Fritzsche, Phys. Rev. Lett. $\underline{37}$, 1504 (1976).
21. M. Kastner, Phys. Rev. Lett. $\underline{28}$, 355 (1972).
22. S. R. Ovshinsky, Phys. Rev. Lett. $\underline{36}$, 1469 (1976).
23. D. Adler, Phys. Rev. Lett. $\underline{41}$, 1755 (1978).
24. D. Adler, J. Non-Crystall. Solids $\underline{35/36}$, 819 (1980).
25. W. E. Spear, in Amorphous and Liquid Semiconductors, edited by J. Stuke and W. Brenig (Taylor and Francis, London, 1974), p. 1.
26. J. C. Knights and G. Lucovsky, CRC Crit. Rev. Solid State Sciences, in press.
27. N. Sol, D. Kaplan, D. Dieumegard, and D. Dubreuil, J. Non-Crystall. Solids $\underline{35/36}$, 291 (1980).
28. M. L. Knotek, Solid State Commun. $\underline{17}$, 1431 (1975).
29. J. C. Knights, J. Non-Crystall. Solids $\underline{35/36}$, 159 (1980).
30. R. C. Frye and D. Adler, to be published.
31. D. L. Staebler and C. R. Wronski, Appl. Phys. Lett. $\underline{31}$, 292 (1977).

Surface Effects and Transport Properties in Thin Films of Hydrogenated Silicon

I. Solomon

Laboratoire P.M.C., Ecole Polytechnique
F-91128 Palaiseau, France

The measurement of transport properties of hydrogenated amorphous silicon, in particular the variation of conductivity with temperature, provides an important test for existing theories [1].

What is predicted is an activated conductivity σ, of the form

$$\sigma = \sigma_o \ \exp(-E_A/kT) \tag{1}$$

where the pre-exponential factor σ_o should be in the range 10^3-$10^4 (\Omega cm)^{-1}$, once the variation of the gap with temperature has been taken into account [2]. The activation energy E_A is the distance of the Fermi level to the mobility edge and is the main parameter governing the transport properties of the material.

1. Band-Bending at Film Surfaces

We consider, in the following, conductivity measurements in the "planar" configuration (Fig.1).

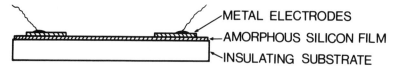

METAL ELECTRODES
AMORPHOUS SILICON FILM
INSULATING SUBSTRATE

Fig.1   Planar configuration for conductivity measurements

The "sandwich" configuration, where the electrodes are on each side of the film, presents difficult electrode problems and the device so obtained has in general diode and carriers-injection properties.

It is easily shown [3] that in the planar configuration, conductivity measurements give results which are extremely sensitive to band-bending at the surfaces of the film. Two parameters will control these band-bending effects (Fig.2) :
- The penetration depth Δ. It is the "Debye length" of the material.
- The value of the band-bending ΔV at the surface.

The basic equations giving the band-bending ΔV and the Debye length Δ in the material are (see for example Ref.[3])

$$\Delta = \frac{1}{e} \sqrt{\frac{\varepsilon \ \varepsilon_o}{\rho}} \tag{2}$$

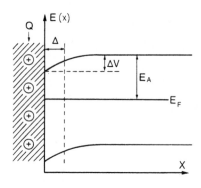

Fig.2 Band-bending at one of the surfaces of a film of an amorphous semiconductor

$$\Delta V = - \frac{Q}{e \sqrt{\epsilon \, \epsilon_o \, \rho}} \qquad (3)$$

where e is the elementary charge, ϵ is the dielectric constant of the material and ϵ_o is the vacuum permittivity, ρ is the density of states of the material near the Fermi level and Q the density of charges at the surface.

The small value of the density of states of hydrogenated amorphous silicon ($\rho \sim 10^{16} cm^{-3} eV^{-1}$) results in a penetration depth Δ of a fraction of a μm, in general of the same order of magnitude than the thickness of the sample. With a conservative value for Q of 10^{10}-10^{11} elementary charges per cm^2, ΔV can be of the order of a few tens of eV. The result is that in most cases the surface conductivity dominates by 3 to 6 orders of magnitude the bulk conductivity of the films.

Thus, unless one uses films of thickness of the order of 1mm (which so far have not been fabricated) in any planar conductivity measurement the results are not in general the bulk values.

2. Experimental Results

We present a rapid review of reported or new experiments where these band-bending effects are clearly demonstrated and play an important role.

2.1. Adsorbed Gases on the Free Surface of the Films

The Chicago Group [4,5] has shown spectacular effects of the adsorption of polar gases (NH_3, H_2O. C_2H_6O, etc) on thin films of hydrogenated silicon. For example variations of 3 to 4 orders of magnitude of the conductance can be obtained by simple exposure of a freshly prepared film to an atmosphere containing a small amount (~2%) of ammonia gas. The original conductance of the film can be partially restored by annealing in vacuum at a temperature of 150°C. A more detailed account of this effect, in particular the time constants involved in these variations of conductance, can be found in the original papers [4,5].

2.2. Variation of the Band-Bending with Temperature

If the conductance of a film is dominated by surface conductivity, one might expect the measured activation energy E_A to be that of the surface.

But this supposes that the band-bending at the surface does not vary with temperature, and we have the evidence that, at least in some cases, this condition is not satisfied. We show, for example, in Fig.3, the variation of the conductance of a film after a temperature step.

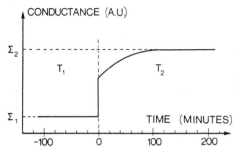

Fig.3 Typical response of the conductance of a silicon film to a temperature step T_2-T_1 at $t = 0$ (Annealed sample of glow discharge silicon in a vacuum of 10^{-7} Torr)

There is a rapid response, followed by a slow variation of the conductance. The rapid part of the variation of the conductance corresponds to an activation energy of about 0.5 eV. For a well annealed sample and in a good vacuum (2×10^{-7} Torr), this fast response is followed by a very slow (several hours) variation of conductance that we interpret as a slow variation of the band-bending at one of the surfaces. At atmospheric pressure or for a non-annealed sample, this variation is much faster and difficult to detect because of the thermal time constant of the apparatus.

If at thermal equilibrium, the band-bending thus varies with temperature, even for a small variation (a few percent) the experimentally measured activation energy E_A and pre-exponential factor σ_o are strongly affected and do not represent any actual values in the material or at the surface of the film.

This can be shown quantitatively in the following equations. For a small variation of the band-bending at the surface with the temperature T around a central temperature T_o, we assume a linear expansion

$$E_A = E_s - \alpha k(T-T_o) +... \qquad (4)$$

Typical values are $T_o = 300°K$, and the actual activation energy at the surface $E_s = 0.5eV$. Taking a rather small variation of the activation energy $\Delta E_A/E_A = 5\%$ for a variation $T-T_o = 40°K$, this gives the value of the dimensionless parameter $\alpha = 7.5$. Now, if we apply the formula of the variable activation energy (4) into the Arrenius equation (1) we obtain at equilibrium

$$\sigma = \sigma_o \exp - \frac{E_o + \alpha kT_o}{kT} + \alpha \qquad (5)$$

So that the experimentally measured quatities are

$$E_A \text{ (apparent)} = E_o + \alpha kT_o \simeq 0.69 \text{ eV} \qquad (6)$$

$$\sigma_o \text{ (apparent)} = \sigma_o \, e^\alpha \simeq 1800 \, \sigma_o \qquad (7)$$

These values are very different from the actual values in the bulk or at the surface. This example shows in particular how it is possible to obtain very large values of the pre-exponential factor σ_0 of the order of 10^6-10^7 $(\Omega cm)^{-1}$ as sometimes reported in the literature and that would be hard to justify theoretically.

2.3. Effect of Light Irradiation on Conductivity

A few years ago, STAEBLER and WRONSKI [6] discovered that films of hydrogenated silicon submitted to visible light irradiation for a few hours could show a spectacular decrease of dark conductance by several orders of magnitude. The initial properties of the sample could be recovered after annealing at a temperature above 150°C. A typical example is reported in Table 1.

Table 1 (After [6])

Sample	Room Temperature Conductivity $[\Omega^{-1}cm^{-1}]$	Activation Energy [eV]	σ_0 $[\Omega^{-1}cm^{-1}]$
Annealed	10^{-6}	0.57	$4 \times 10^{+3}$
Irradiated	2×10^{-10}	0.87	10^5

The exact origin of the effect is not quite established. However, the following experiment shows that it is due, at least partially, to a variation of surface conductivity, most reasonably due to a band-bending effect (Fig.4). We consider a comparatively thick sample (5.9 μm) which in the annealed state has a resistance $R_{DARK} = 10^{11}\Omega$. We irradiate the film with "blue light" (λ = 450 nm) during 4 hours and the measured resistance of the film then becomes $R'_{DARK} = 10^{12}\Omega$. Since the irradiating light penetrates only 1/60 of the sample, the blue irradiation can only affect the surface of the sample (Except for a very unlikely diffusion of the created carriers accross the thick sample). Since the resistance of the sample has increased under irradiation, one can only conclude that in the annealed state the conductance was dominated, at least for 90%, by surface conductivity. Depending upon the sample, the effect is larger by irradiation through the substrate as in the figure, or on the free surface of the sample.

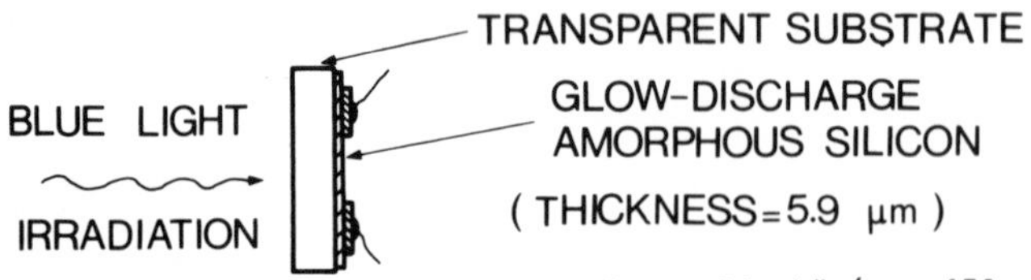

Fig.4 Experimental set-up. The "blue ligth" (λ = 450 nm) penetrates only 0.1 μm into the sample

Our model of the decrease of the band-bending by light irradiation can be explained qualitatively in the following manner. As it is well known, an intense light irradiation, creating a large number of mobile carriers, tends to decrease any built-in electric field and thus to suppress any band-bending in the material. If some of the carriers thus accumulated at different distance of the surface are trapped, some memory of the charge distribution under illumination will be retained after suppression of the irradiating light. The net effect will be a decrease, after prolongated light irradiation, of the band-bending resulting, in extreme cases, in a "flat band" condition even in the dark.

36

2.4. Quenching of Luminescence by Surface Effects

In a very recent work, D. DUNSTAN [7] has shown that the luminescence properties of the surface of hydrogenated amorphous silicon at liquid nitrogen
temperature are different for an annealed sample or for a sample that has
been irradiated for a long time (as for the Staebler-Wronski effect described in 2.3).

The experimental arrangement, shown schematically in Fig.5, is made of :
- A modulated exciting light, that can be "blue" (λ = 450 nm) or "red"
(λ = 650 nm), and the corresponding modulated luminescence is coherently
detected with a lock-in amplifier.
- An intense source of continuous white light ("bias illumination") which
is not detected directly by the luminescence detector since it is not modulated.

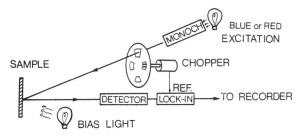

Fig.5 Experimental set-up for the detection of surface effects on luminescence. During irradiation with the "bias light", the large number of photo-
created carriers suppress the built-in electric field due to band-bending
at the surface of the sample

What is observed, in this experiment, is the variation of the luminescence intensity upon application of the white bias illumination. The following result is obtained for very low modulated exciting light (10^{12} photons
/cm^2 s or less)

Table 2

	Blue Exc.	Red Exc.
Irradiated sample	no effect	no effect
Annealed sample	increases	no effect

Our explanation of these results are based on the fact that luminescence
in this type of material decreases upon the application of an electric
field ("electric field quenching of luminescence") [8]. We believe that in
the present experiment, the luminescence is slightly quenched by the
built-in electric field due to the band-bending at the surface (For a
band-bending of 0.2 eV and a Debye length of about 0.2 μm, the built-in
electric-field is of the order of 10^4 V/cm). The effect of the white bias
excitation is to create a large number of carriers which will suppress the

built-in electric field, as discussed in 2.3. This cancels the effect of
the electric field quenching and thus results in <u>an increase</u> of the luminescence intensity. From the results of Table 2, we see that this is obtained when we have simultaneous conditions : a) blue irradiation with a low
penetration depth ($\simeq 0.1 \mu$m) so that it excites the luminescence near the
surface only ; b) annealed sample which according to our discussion of 2.3
has a surface band-bending, whereas an irradiated sample is near the flat
band condition and thus gives no effect. We remark that the red excitation,
because it measures the luminescence in the bulk of the 3 μm-thick sample,
shows no effect. Of course, these effects can be observed only for very low
excitations, otherwise the exciting light itself would have the same effect
as the bias light and would suppress the band-bending to be observed
(Fig.6).This results in a rather low luminescence signal which is the main
difficulty of the experiment. However, it is potentially a very direct method to measure the band-bending at the surfaces of a film. In particular,
for the samples observed, D. DUNSTAN has found that the band-bending at the
silicon-substrate interface is larger than at the free surface of the film
(Fig.6). And also, by inhomogeneous light irradiation, he has been able to
obtain different values of band-bending along the surface of the film [7].

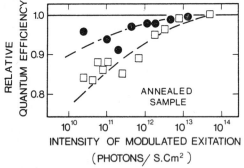

Fig.6 Variation, with the intensity of the "blue" excitation (λ = 450 nm),
of the quantum efficiency of luminescence at the surface of an annealed
film of glow-discharge amorphous silicon (□ □ □ excitation through the
transparent substrate. ● ● ● free surface). (Liquid nitrogen temperature).
There is no effect in the case of red light excitation or with a film which
has been irradiated by a high intensity light for a few hours (After [7]).
The quantum efficiency with the "bias light" is taken as reference

Very recently also, the Marburg group [9] have demonstrated the existence of surface and thickness effects in hydrogenated amorphous silicon.
They have found that the decay time of luminescence is much shorter at the
surfaces of the films that in the bulk. As in [7], they explain these surface effects by the presence of an electric-field at the surfaces of the
samples.

3. Conclusion

It is thus extremely difficult, if possible at all, to measure bulk conductivity in films of undoped hydrogenated amorphous silicon. One can try to
control more closely one of the surfaces of the film [3,4] and to interpret

the results as bulk properties [10]. Unfortunately, there is always a doubt about the effect of the other surface, and this is also a problem in field-effect measurements [11].

The same kind of difficulties appears in photoconductivity measurements [6,12], and surface effects are found even in films of doped material [13,14] although the effect of the band-bending is quantitatively much smaller.

It is thus a paradoxical conclusion to realize that, after a few years of active studies, not so much is known about bulk transport and photoconductivity properties of hydrogenated amorphous silicon. This is rather surprising for a material of economic importance, which is one of the best candidates for the fabrication of inexpensive solar cells.

References

1. N.F. Mott, Phil. Mag. 26, 1015 (1972).
2. J. Perrin and I. Solomon, J. of Non-Crystal. Sol. 37, 407 (1980).
3. I. Solomon, T. Dietl and D. Kaplan, Le Journal de Physique 39, 1241 (1978).
4. M. Tanelian, H. Fritzsche, C.C. Tsai and E. Symbalisty, Appl. Phys. Lett. 33, 353 (1978).
5. M. Tanelian, M. Chatani, H. Fritzsche, V. Smid and P.D. Persans, J.of Non-Cryst. Sol. 35 and 36, 575 (1980).
6. D.L. Staebler and C.R. Wronski, Appl. Phys. Lett. 31, 292 (1977).
7. D. Dunstan, to be published.
8. W. Rhem and R. Fischer, Phys. Stat. Sol.(b) 94, 595 (1979).
9. W. Rhem, R. Fischer and J. Beichler, to be published.
10. I. Solomon, F. Perrin and B. Bourdon, 14th Int. Conf. Phys. Semicond. (Edinburgh 1978), Published in Inst. Phys. Conf. Ser. N° 43 (1979) p. 689.
11. N.B. Goodman, H. Fritzsche and H. Ozaki, J. of Non-Cryst. Sol. 35 and 36, 599 (1980).
12. P.D. Persans and H. Fritzsche, A.P.S. Bull. 24, 399 (1979).
13. D.G. Ast and M.H. Brodsky, Phil. Mag. 1980, in press.
14. I. Solomon and M.H. Brodsky, J. Appl. Phys. 1980, in press.

The Past, Present and Future of Amorphous Silicon

W.E. Spear

Carnegie Laboratory of Physics, University of Dundee
Dundee DD1 4HN, Scotland, U.K.

During the last few years interest in both fundamental and applied aspects of amorphous semiconductors has grown rapidly. In the following, which is a summary of my talk at the 1980 Kyoto Summer Institute, I should like to trace the development of the subject and then go on to discuss some of the recent work in our laboratory. The second part will be continued in the talk by Dr. P.G. Le Comber.

1. The Past

1.1 Development and Properties of Glow Discharge Silicon

H. STERLING and his collaborators at the S.T.L. Laboratories in Harlow, Essex, were probably the first to prepare thin films of a-Si and a-Ge by the decomposition of the hydride in a r.f. glow discharge. The work started in the mid-sixties and, although it was discontinued in 1969, led to several interesting papers on the preparation technique and the properties of the films [1,2,3].

The Dundee Group entered the a-semiconductor field towards the end of 1968. We were stimulated by the new concepts and ideas that had been introduced by Mott and others and the initial aim of our work was to provide some reasonably conclusive experimental tests of the proposed models and transport mechanisms. After investigating evaporated, sputtered and glow discharge films of a-Si and a-Ge, we concluded at an early stage of the work that the glow-discharge technique was the most promising approach for our purpose; it led to material in which the basic properties of the a-phase were not obscured by a high density of defect states. For this reason, since 1969, we have concentrated our efforts on the development of the glow discharge method for the preparation of a-Si, a-Ge and other materials.

In 1970 we published the first electron drift mobility results [4] on glow discharge a-Si. The temperature dependence of drift mobility and conductivity led to the conclusion that above about 250K electrons propagate in the extended states with a mobility between 1 and 10 $cm^2V^{-1}s^{-1}$. Below that temperature, phonon-assisted hopping through localised states near the bottom of the band tail begins to predominate and eventually, with decreasing temperature, the conduction path moves towards the Fermi level. This interpretation, which supported some of Mott's basic ideas, appears to have stood the test of time.

We believed at this stage, that the widely different electronic properties found in evaporated and glow discharge a-Si specimens were produced by different densities and distributions of localised states in the mobility gap. Experimental information on the gap state distribution was evidently of considerable importance and we began to develop a field effect technique for this purpose. The first, somewhat tentative results [5] were reported at the 1971 Ann Arbor conference and more detailed information [6] on the density of states distribution was given at the 1973 Garmisch Meeting and in subsequent papers [7,8]. During 1973 and 1974 a fairly detailed study of the optical and photoconductive properties of glow discharge Si [9,10] was carried out at Dundee, in which it was attempted to correlate the photoconductive behaviour with transport results and the density of state distribution.

The most striking feature brought out by the early field effect work was the remarkably low overall density of gap states that could be achieved in glow discharge a-Si deposited at substrate temperatures around 250°C. $g(\epsilon)$ in the centre of the gap appeared to be at least two orders of magnitude lower than in evaporated or sputtered a-Si. It occurred to us that the negative results of previous doping experiments on these materials could be primarily caused by the large $g(\epsilon)$ rather than by the fundamental impossibility of producing donor or acceptor configurations in the a-phase. In 1975 LE COMBER and I showed [11,12] that the electronic properties of a-Si and a-Ge could be controlled by substitutional doping in a systematic way over a remarkably wide range. This was followed by the first thin film a-Si p-n junction [13] and shortly afterwards by a paper on the independent development of a-Si photovoltaic devices in the RCA Laboratories [14].

The possibility of substitutional doping has removed one of the main limitations and opened up an exciting new field for fundamental and applied developments. Since 1976 many research groups in Europe, the United States and Japan have began work on plasma deposited a-Si and it becomes difficult at this stage to extend the summary further. The remarkable growth of interest in plasma deposited a-Si is reflected in the number of papers read on the subject at the biennial International Conferences on Amorphous and Liquid Semiconductors. In 1971 (a-4) and 1973 (a-5) the only contributions came from the Dundee group, but in 1977 (a-7) the number had risen to 24 and in 1979 (a-8) to about 80.

1.2 Work on Sputtered and Evaporated a-Si and a-Ge

During the period from 1969 to 1976, discussed above, a considerable amount of work was carried out on a-semiconductors prepared by sputtering and evaporation, which has made important contributions to the development of the subject. I should like to mention in particular the information on the nature and structure of the defects in tetrahedrally co-ordinated a-materials obtained at the IBM laboratories by BRODSKY and his colleagues [15], at Harvard University by PAUL'S group [16] and at Marburg by STUKE and his collaborators [17].

It had been suggested as early as 1970 [15] that one of the reasons for the promising properties of glow discharge a-Si might be the presence of hydrogen in the plasma near the specimen surface which saturated dangling bonds during growth. An important development in this direction was the work of LEWIS et al at Harvard in 1974 on the hydrogenation of sputtered a-Ge films [18]. In 1976 PAUL and his colleagues were able to show [19] that the addition of hydrogen to the argon sputtering gas has a marked effect on the electronic properties of the sputtered a-Si film. The

material can now be doped with reasonable efficiency by the addition of
phosphine (or diborane) and this result has contributed to the present applied interest in a–Si.

The role of hydrogen in a–semiconductors has stimulated a great deal of
experimental work to determine its concentration, distribution and configuration in the deposited specimen. Since about 1975 techniques such as
thermal evolution [20], the absorption in the Si–H vibrational spectrum
[21] and the nuclear resonance reaction [22] have been applied to the hydrogen problem. A detailed review of hydrogen in a–semiconductors is being
published by KNIGHTS and LUCOVSKY [23].

2. The Present

Under this heading I should like briefly to discuss two projects under
investigation in our laboratory. The first concerns the temperature
dependence of the reference energies ϵ_f and ϵ_c and is of basic importance to the interpretation of transport results. In recent papers [24,
25,26], we considered this problem in some detail. As far as the Fermi
level is concerned, there are two mechanisms which can cause a thermal
shift of ϵ_f with respect to the onset of the tail states at ϵ_A. The
first is the electron-phonon interaction, which, as in crystalline Si, also
narrows the mobility gap in the a-phase with increasing T [27]. Experimental evidence in undoped a-Si points towards a temperature coefficient
$\delta_f \simeq 2 \times 10^{-4}$ eV K^{-1}, such that $(\epsilon_A - \epsilon_f)$ decreases with rising T. In
n-type specimens, where ϵ_f may lie in regions of rapidly varying $g(\epsilon)$,
the statistical shift of ϵ_f [12] will result in a movement opposite to
that of the electron-phonon interaction, but having approximately the same
magnitude of $|\delta_f|$.

The experimental results, particularly those in ref. [26], suggest that
in undoped a-Si the temperature coefficient associated with $\epsilon_c(T) - \epsilon_f(T)$
approaches 10^{-3} eV K^{-1}, several times the value of δ_f. This implies that
there has to be an appreciable movement of the current path at ϵ_c towards
ϵ_A as T is increased. The explanation of this effect is still a controversial matter. In ref. [24] an attempt has been made to interpret it in
terms of the increase with temperature in the average wavefunction overlap
between electron states at ϵ_c. Although the model is based on a highly
simplified calculation it nevertheless has proved successful in the consistent interpretation of a number of experimental results. The predicted
proportionality between δ_c and the tail state width $\epsilon_c(0) - \epsilon_A$ introduces
a new criterion for explaining the variations in the pre-exponential term
σ_0 of the conductivity expression. Also, the widely observed high-temperature discontinuity in the conductivity activation energy of undoped a-
Si can be understood in terms of the movement of the current path with increasing T towards a limiting position near ϵ_A.

Further evidence has been obtained from 'photoconductive probing' experiments [26] in the weakly absorbed spectral region between 0.45 and 1.1 eV.
The aim has been to determine the threshold energies $\Delta\epsilon_1$ and $\Delta\epsilon_2$ for
transitions from localised initial states just below ϵ_f into final extended states at ϵ_c and final localised tail states at ϵ_A respectively. In
fig.1 the threshold values are plotted as a function of temperature for an
undoped a-Si specimen. The linear part of the $\Delta\epsilon_1$ curve leads to a total temperature coefficient $\delta = \delta_c + \delta_f$ of 9×10^{-4} eV K^{-1}, whereas the
$\Delta\epsilon_2$ curve gives $\delta_f \simeq 2 \times 10^{-4}$ eV K^{-1}. On the high temperature side

extrapolation defines a limiting temperature $T^{\dagger}$ at which the current path
approaches ϵ_A. It is encouraging that the above mentioned high temperature discontinuity in the dark conductivity is also observed at a temperature close to $T^{\dagger}$.

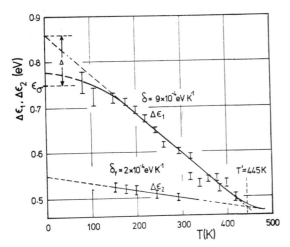

Fig.1   Threshold values $\Delta\epsilon_1$ and $\Delta\epsilon_2$ for transitions to ϵ_c and ϵ_A
plotted as a function of temperature . (From ref. 26)

The second subject is an applied one, concerned with the development of
a-Si junctions of high current carrying capacity.   As described in a recent paper [28], various junction profiles were investigated, both smoothly
graded and abrupt.   We found that the n^+-ν-p^+ configuration gave the most
promising results.   The devices were deposited by the glow discharge technique on to stainless steel substrates coated with a thin film of chromium.
The n^+ layer, a few hundred A° thick, was followed by a lightly phos-
phorous-doped ν region ($\simeq 1\mu$m) and a thin p^+ layer.   An evaporated gold
top contact was used.

Figure 2 shows a series of characteristics for such junctions plotted
on a semi-logarithmic graph of the current density J against the applied
potential V.   The current carrying capacity in the forward direction increases with the doping level of the central region and reaches its optimum
when about 40 vppm of phosphine are added to the silane.   Steady current
densities of 20 A cm^{-2} or more are obtainable.   The rectification ratio at
V = 1V, the approximate barrier height, is about 5 x 10^4.   The reverse
characteristics show reversible breakdown phenomena which are possibly associated with the Zener mechanism.   The diode quality factor has been determined from the J-V characteristics and from measurements of the open-
circuit voltage under illumination.   Both lead to values of about 1.5.

The results are encouraging and suggest that thin film a-Si diode structures could provide a cheap alternative to crystalline junctions in a range
of less demanding applications where the characteristics of Fig.2 would be
adequate.

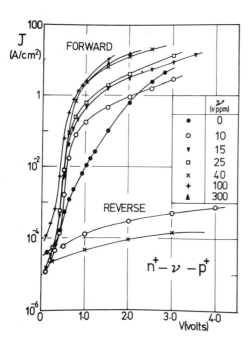

Fig.2 J-V characteristics of n⁺-ν-p⁺ diodes at different doping levels for the central ν region. (From ref. 28)

3. The Future

As pointed out in 1977 [29], future support for fundamental developments in the a-semiconductor field will depend to a large extent on the progress that is being made on the applied side. One of the major problems is the accurate control of the glow discharge plasma, necessary for reproducible and optimised electronic specimen properties. I was impressed by my recent visits to Japanese industrial and University laboratories, where different aspects of the plasma problem are being investigated with some success. I am therefore encouraged in the belief that a-semiconductors, and a-Si in particular, will continue to play a significant role in both fundamental and applied developments.

References

1. H.F. Sterling, R.C.G. Swann: Solid State Electronics, $\underline{8}$, 653-654 (1965)

2. R.C. Chittick, J.H. Alexander, H.F. Sterling: J.Electrochem.Soc. $\underline{116}$, 77-81 (1969)

3. R.C. Chittick: J. Non-Cryst. Solids, $\underline{3}$, 255-270 (1970)

4. P.G. Le Comber, W.E. Spear: Phys. Rev. Letters, $\underline{25}$, 509-511 (1970)

5. W.E. Spear, P.G. Le Comber: J. Non-Cryst. Solids, $\underline{8-10}$, 727-738 (1972)

6. W.E. Spear: Proc. of 5th International Conference on Amorphous and Liquid Semiconductors, ed. by J. Stuke and W. Brenig, (Taylor and Francis) (1974) pp 1-16

7. A. Madan, P.G. Le Comber, W.E. Spear: J. Non-Cryst. Solids, 20, 239-257 (1976)

8. A. Madan, P.G. Le Comber: Proc. of 7th International Conference on Amorphous and Liquid Semiconductors, Edinburgh, ed. W.E. Spear (CICL, Univ. of Edinburgh), (1977) pp 377-381

9. R.J. Loveland, W.E. Spear, A. Al-Sharbaty: J. Non-Cryst. Solids, 13, 55-68 (1973/74)

10. W.E. Spear, R.J. Loveland, A. Al-Sharbaty: J. Non-Cryst. Solids, 15, 410-422 (1974)

11. W.E. Spear, P.G. Le Comber: Solid State Comm. 17, 1193-1196 (1975)

12. W.E. Spear, P.G. Le Comber: Phil. Mag. 33, 935-949 (1976)

13. W.E. Spear, P.G. Le Comber, S. Kinmond, M.H. Brodsky: Appl. Phys. Letters, 28, 105-107 (1976)

14. D.E. Carlson, C.R. Wronski: Appl. Phys. Lett. 29, 602-605 (1976)

15. M.H. Brodsky, R.S. Title, K. Weiser, G.D. Pettit: Phys. Rev. B1, 2632 (1970)

16. N.J. Shevchik, W. Paul: J. Non-Cryst. Solids, 8-10, 381-387 (1972)

17. W. Beyer, J. Stuke: ref. 6, 251-258 (1974), see also Phys. Stat. Sol. (a) 30, 511-520 (1975)

18. A.J. Lewis, G.A.N. Connell, W. Paul, J.R. Pawlik, R.J. Temkin: Tetrahedrally Bonded Amorphous Semiconductors, ed. by M. Brodsky, S. Kirkpatrick and D. Weaire (A.I.P. New York, 1974) pp. 27-32

19. W. Paul, A.J. Lewis, G.A.N. Connell, T.D. Moustakas: Solid State Comm. 20, 969-972 (1976)

20. A. Triska, D. Dennison, H. Fritzsche: Bull. APS 20, 392 (1975)

21. M.H. Brodsky, M. Cardona, J.J. Cuomo: Phys. Rev. B. 16, 3556-3571 (1977)

22. M.H. Brodsky, M.A. Frisch, J.F. Ziegler, W.A. Lanford: Appl. Phys. Lett. 30, 561-563 (1977)

23. J.C. Knights, G. Lucovsky: To be published in 'Critical Reviews of Solid State Sciences'.

24. W.E. Spear, D. Allan, P.G. Le Comber, A. Ghaith: Phil. Mag. B. 41, 419-438 (1980)

25. W.E. Spear, D. Allan, P.G. Le Comber, A. Ghaith: J. Non-Cryst. Solids, 35 and 36, 357-362 (1980)

26. W.E. Spear, H. Al-Ani, P.G. Le Comber: Phil. Mag. 41 (1980) in press

27. R.W. Griffith: J. Non-Cryst. Solids, 24, 413 (1977)

28. R.A. Gibson, P.G. Le Comber, W.E. Spear: Appl. Phys. 21, 307-311 (1980)

29. W.E. Spear: Adv. in Phys. 26, 811-845 (1977)

Doping and the Density of States of Amorphous Silicon

P.G. LeComber

Carnegie Laboratory of Physics, University of Dundee
Dundee DD1 4HN, Scotland, U.K.

1. Introduction

In the preceding article, Professor Spear has given a brief historical survey of the work that led to the successful doping of a-Si and to the present
interest in its application in solar cells, high current diodes, liquid
crystal displays, etc., and has discussed a number of fundamental topics of
current interest.   In this article the methods that have been used to dope
a-Si are described.   The main reason why a-Si can be doped is its low density of gap states.   Probably the most complete information on the density
of states function $g(\mathcal{E})$ comes from the extensive field effect measurements
of the Dundee group.   Recently a number of papers have raised doubts about
the analysis used to calculate $g(\mathcal{E})$.   We therefore review the results of
bulk measurements that provide support or otherwise for the features of $g(\mathcal{E})$
determined from the field effect experiments.   Finally the experimental
approach used in the field effect experiments has been applied in the development of an insulated-gate field-effect transistor and we briefly review the
present stage of this work.

2. Preparation of Doped Amorphous Semiconductors

A review of the methods by which doped amorphous semiconductors have been
prepared has recently been published [1] and therefore only a brief discussion will be given here.   Essentially five techniques have been used to date,
namely glow discharge deposition, rf sputtering, ion implantation, thermal
diffusion and post-hydrogenated CVD, and in the following we shall consider
each of these in turn.

2.1 Doping by the Glow Discharge Technique

This technique has received a considerable amount of attention in recent
years since it was used in the first successful doping experiments on amorphous Si and Ge by SPEAR and LE COMBER in 1975 [2,3].   Thin film specimens
of a-Si or a-Ge were formed by decomposing the corresponding hydride, silane
(SiH_4) or germane (GeH_4), in a radio frequency field.   The power level is
small, typically a few watts, and frequencies between 1 and 100MHz have been
tried.   The apparent simplicity of the technique is deceptive.   The electronic properties of the deposited specimens are critically dependent on a
number of variables such as the temperature of the substrate during deposition [4], flow rate, gas pressure, rf power level, the floating potentials
on specimen and other surfaces, and the geometry of the reaction tube.

Doping from the gas phase can be achieved by adding small but accurately determined amounts of phosphine (PH_3) or diborane (B_2H_6) to the silane. The room temperature conductivity σ_{RT} of a-Si prepared in this way is shown in Fig.1 as a function of the gaseous impurity ratio [3].

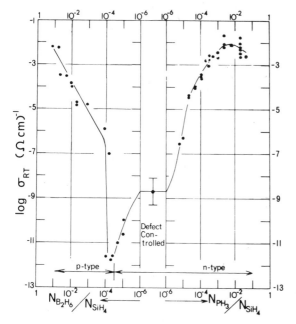

<u>Fig.1</u> Room temperature conductivity σ_{RT} of n- and p- type a-Si specimens plotted as a function of the impurity ratios, PH_3/SiH_4 and B_2H_6/SiH_4 respectively, used in the gas mixture for specimen preparation. The centre refers to undoped specimens (After ref.3)

On the right-hand side this is the ratio of the number of phosphine to the number of silane molecules in the gaseous mixture, whereas on the left the corresponding diborane to silane ratio is shown. In the centre of the graph, the conductivities of 10^{-8} to 10^{-9} $\Omega^{-1}cm^{-1}$ are representative of undoped glow discharge specimens. These results demonstrate clearly that σ_{RT} can be controlled over some ten orders of magnitude, from 10^{-12} to 10^{-2} $\Omega^{-1}cm^{-1}$, with both n- and p- type doping, as the Fermi level is moved over approximately 1.2eV [3].

Subsequently, MADAN et al [5] published similar results for specimens prepared from gas mixtures of SiF_4 and H_2. They obtained σ_{RT} values appreciably higher than unity for very small concentrations of PH_3 or AsH_3 in their preparation mixtures. Two pieces of experimental evidence have recently come to light which cast serious doubts on their claims that these results on Si:F:H films suggested a much lower $g(\mathcal{E})$ than in silane produced a-Si:H. Firstly, MATSUDA et al [6] have shown that doped films produced under the high-power conditions necessary for deposition in the SiF_4 mixtures are polycrystalline rather than amorphous. Secondly, for a given concentration of doping gas in the initial mixture, the amount of atomic impurity incorporated in the Si:F:H films is over one hundred times larger

than in the silane produced films [3,6,7].   Clearly, any comparison of the
doping efficiency and the density of states in these materials is without
foundation since one material is polycrystalline and one is amorphous.
Furthermore one should really compare doping efficiencies in terms of the
number of impurities incorporated in the film rather than in terms of gaseous impurity ratios in the mixtures used to prepare the films.   More
information is needed before valued judgements can be made on the relative
merits of films produced from silane or from SiF_4+H_2 mixtures.

2.2 Hydrogenated Sputtered Amorphous Semiconductors

The work of the Harvard group on hydrogenated sputtered a-Ge and Si has made
an important contribution to the subject.   It is generally accepted that
specimens prepared by cathodic sputtering contain a high overall level of
gap states associated with dangling bonds in probably quite an extensive
microvoid structure.   PAUL and his collaborators [8,9] showed that by adding hydrogen to the argon used in the sputtering process, a few atomic percent of hydrogen can be incorporated into the films.   This has a marked
effect on the electronic properties of the material, greatly reduces the ESR
signal [10], and is consistent with the suggestion that dangling bonds have
been saturated by hydrogen atoms.   In the present context, the significant
result is that fairly sensitive doping from the gas phase using phosphine
and diborane now becomes possible [11].

2.3 Doping by Ion Implantation

Doping by ion implantation has achieved considerable importance in crystalline device technology.   Although earlier attempts with evaporated a-Ge were
not encouraging, recent work on glow discharge a-Si has shown that it is
possible to dope this material by implantation of the substitutional impurities P, As, Sb, Bi, B, Al, Ga, In, Tl and the interstitial impurities Na,
K, Rb and Cs [12,13,14,15].   Implantation doping provides the same range
of control of the electrical properties as gas-phase glow discharge doping,
but at a lower doping efficiency.   Compensation of pre-doped specimens by
ion implantation has been investigated, found to be feasible and predictable,
and used for the production of implanted a-Si p-n junctions.

2.4 Doping by Thermal Diffusion

BEYER and FISCHER [16] have doped glow discharge a-Si interstitially with
Li both by thermal diffusion and by ion implantation.   For the former, a
layer of Li was evaporated onto the undoped material and then heated to
$400^{\circ}C$ for one hour to allow the Li to diffuse into the sample.   Under
these conditions the diffusion length is likely to be greater than the sample
thickness so that a homogeneous distribution is obtained.

2.5 Post-hydrogenation of CVD Films

Recently two research groups [17,18] have doped a-Si films prepared by
chemical vapour deposition (CVD).   The films are prepared by the thermal
decomposition of mixtures of silane, hydrogen and impurity gases at a
temperature close to $600^{\circ}C$.   SOL et al [17] then annealed the films between
400 and $500^{\circ}C$ in a hydrogen plasma.   The technique is not easy since the
annealing stage can result in plasma etching and loss of the film!   Furthermore it is difficult to diffuse into the a-Si a uniform hydrogen concentration of sufficient magnitude to produce a material with good electronic
properties.   Nevertheless, since room temperature conductivities approach-

ing 0.7 Ω^{-1}cm^{-1} can be obtained in this way, the technique is clearly capable of producing doped a–Si.

2.6 Doping and the Density of States

Efficient doping of a–Si is only possible for material prepared by the glow discharge method, for hydrogenated sputtered films or for post-hydrogenated CVD material. Why? The answer to this question has been discussed previously [1,3] and has to do with the lower density of localized states, particularly near the Fermi energy, of these materials compared to, for example, evaporated a–Si. In the next section we discuss this important property in more detail.

3. Field Effect and the Density of States in a–Si

The density of states function of an amorphous material is one of the most important parameters that can be used to characterise the material. As such the experimental determination of $g(\mathcal{E})$ is of considerable importance. In the preceeding article Professor Spear has outlined the development at Dundee of the field effect technique for the determination of $g(\mathcal{E})$ in a–Si. In spite of the many difficulties in the analysis of the data, this technique still provides the most extensive information on $g(\mathcal{E})$ available to date. In the following we shall therefore describe the results obtained and, in view of recent criticisms, look for corroboration of the main features from independent measurements of bulk properties.

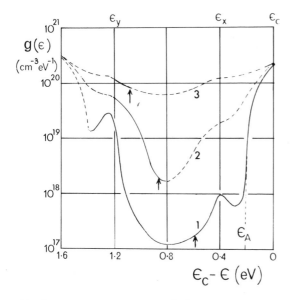

Fig.2 Density of states $g(\mathcal{E})$ in a–Si determined from field effect data. Curve 1: glow discharge sample deposited at about 550K; curve 2: glow discharge sample deposited at about 350K; curve 3: evaporated or sputtered film. (After ref.20)

49

The field effect techniques and the method of analysis have been discussed extensively in the Dundee publications [19,20,21]. The results obtained for glow discharge a-Si specimens deposited onto substrates held at a temperature of about 550 K (curve 1) are summarized in Fig.2. The general features of the results are the low overall density of gap states, the broad minimum in $g(\mathcal{E})$ near the centre of the mobility gap, the peaks at $\mathcal{E}_x$ and $\mathcal{E}_y$ and the rapid rise in $g(\mathcal{E})$ at $\mathcal{E}_A$ corresponding to the onset of the electron tail states. GOODMAN and FRITZSCHE [22] and POWELL [23] have suggested in a recent paper that some of these features may be a consequence of the step-by-step analysis used by SPEAR and LE COMBER [19]. In contrast GRUNEWALD et al [24] have developed a method of calculating $g(\mathcal{E})$ that does reproduce the features [25] obtained by Spear and Le Comber to within a factor of two or three, which is about the uncertainty the Dundee group have estimated in their $g(\mathcal{E})$ values. Clearly the field effect is not sufficiently accurate to measure small features in $g(\mathcal{E})$, such as the peak at $\mathcal{E}_x$, with any certainty and furthermore if the density of states in the surface is different from that in the bulk then further errors are introduced. As stressed in our previous publications [19,21], it is essential to provide independent evidence for the features in $g(\mathcal{E})$ calculated from the field effect data. In view of the importance of $g(\mathcal{E})$ in determining many of the properties of a-Si we have therefore summarized the data that can be used as an independent check of the field effect $g(\mathcal{E})$ data shown in Fig.2.

3.1 Peak at $\mathcal{E}_y$

'(a) Optical absorption and photoconductive spectra have a feature at $\hbar\omega$ of 1.1 to 1.2eV which could correspond to transitions from $\mathcal{E}_y$ to $\mathcal{E}_c$ [26].
(b) It becomes increasingly more difficult to shift $\mathcal{E}_F$ with boron doping as $\mathcal{E}_F$ is moved into the region of $\mathcal{E}_y$ by either gas phase doping [3] or by ion implantation [13,14,15] consistent with a large density of states at $\mathcal{E}_y$.
(c) Hole drift mobility data leads to the conclusion that the transport of excess holes is controlled through trapping and thermal release by states about 0.45eV above $\mathcal{E}_v$, i.e. at $\mathcal{E}_y$ [27].
(d) Measurements of barrier capacitance are consistent with the rapid increase in $g(\mathcal{E})$ at $\mathcal{E}_y$ [28].
(e) Conduction in glow discharge films deposited at a low temperature changes from $\mathcal{E}_c$ conduction to holes hopping through a feature in $g(\mathcal{E})$ about 1.1eV below $\mathcal{E}_c$, i.e. at $\mathcal{E}_y$ [4]. The magnitude of this hole hopping path, i.e. magnitude of $g(\mathcal{E}_y)$, increases as the deposition temperature is decreased. (see also section 3.5 below).

3.2 Peak at $\mathcal{E}_x$

This is probably the most difficult feature to verify because of its relatively small height in material with good electronic properties. However, some evidence for its existence can be obtained from:-
(a) the relative $g(\mathcal{E})$ calculated from the shift of $\mathcal{E}_F$ with ion implantation doping (see section 3.6 below).
(b) Photoconductivity measurements in doped a-Si which suggest that a transition to a bi-molecular recombination mechanism takes place when the quasi-Fermi level has moved through the $\mathcal{E}_x$ maximum, producing a reservoir of initial states for the recombination process [29].

3.3 Rapid Rise in g($\mathcal{E}$) above $\mathcal{E}_A$

(a) Interpretation of electron drift mobility data [4,30] in terms of trap-controlled extended state transport. The drift mobility data located the onset of the tail states about 0.2eV below $\mathcal{E}_c$ in excellent agreement with the g($\mathcal{E}$) data.
(b) It becomes extremely difficult to move $\mathcal{E}_F$ much closer than about 0.18eV from $\mathcal{E}_c$ with n-type gas phase doping [3] or by ion implantation [13,14,15]. This is consistent with the rapid rise in g($\mathcal{E}$) observed at $\mathcal{E}_A$ in the field effect measurements.

3.4 Broad Minimum in g($\mathcal{E}$) near the Centre of the Gap

(a) It is relatively easy to shift $\mathcal{E}_F$ by gas phase doping [3] or ion implantation [13,14,15] when $\mathcal{E}_F$ is near the centre of the gap – therefore relatively low g($\mathcal{E}$) in this region.
(b) Recent results in our laboratory [31] show that the g($\mathcal{E}$) values measured from field effect increase as the samples are annealed to remove hydrogen [32]. This is consistent with the observed increase in the hopping component of the dark conductivity and with the observed increase in the monomolecular recombination of photoexcited carriers, both of which are believed to take place via states at $\mathcal{E}_F$.

3.5 Effect of Deposition Temperature on g($\mathcal{E}$)

The early work of LE COMBER et al [4] suggested, on the basis of conductivity data, that the overall g($\mathcal{E}$) increased as the deposition temperature T_d of the films was lowered. Subsequent field effect measurements [20] provide support for this suggestion. Conversely, the conductivity results may also be said to provide an independent check on these field effect results. Further support for an increase in the overall g($\mathcal{E}$) as T_d is decreased are obtained from –
(a) Photoconductive lifetime decreases as T_d decreases [25].
(b) Luminescence efficiency decreases as T_d decreases [33].
(c) Electron spin density increases as T_d decreases [34].
All of which are consistent with an increasing g($\mathcal{E}$) as T_d is lowered.

3.6 Relative g($\mathcal{E}$) from Shift of $\mathcal{E}_F$ by Ion Implantation

If we make two simplifying assumptions [14] it is possible to estimate the relative g($\mathcal{E}$) from the number of implanted ions required to shift $\mathcal{E}_F$ by a small amount. This approach has been used by BEYER et al [35] using Li implantations and they find good agreement with the shape of g($\mathcal{E}$) between about 0.65eV and 0.15eV below $\mathcal{E}_c$, including the feature at $\mathcal{E}_x$. KALBITZER et al [14] have similarly used alkali (A), P and B implantations to obtain the results shown in Fig.3 which are compared with the field effect results shown by the broken line. Clearly these results also support the features in g($\mathcal{E}$) deduced by the field effect technique.

3.7 Conclusions

In conclusion, it is apparent that these results, together with the above evidence, provide convincing support for the general features of the density of states deduced from field effect experiments. Although the absolute g($\mathcal{E}$) may not be accurate to within a factor of two or three it appears that in spite of the computational difficulties associated with the analysis of the experimental data, the field effect experiment provides us in a-Si with

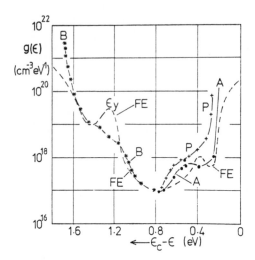

Fig.3 Approximate density of states distribution of a-Si calculated from the results of phosphorus, boron and alkali implantations (curves P,B and A). The field effect data (FE) are shown for comparison.   Curves P,B and A are normalized to the minimum of FE.   (after Ref.14)

a $g(\mathcal{E})$ distribution that can be correlated with many independent measurements of bulk properties.

4. Application of Field Effect Techniques to a Thin Film Transistor

Early on in the development at Dundee of the field effect techniques, it became apparent that the low $g(\mathcal{E})$ of a-Si, and the resulting rapid changes in source-drain current with gate voltage, provided the basis for a thin film transistor in the form of an insulated-gate field-effect transistor or igfet [36].   Subsequent collaboration between our laboratory and RSRE at Malvern has demonstrated that these devices are particularly suited [37] to the active elements of an X-Y addressable matrix for liquid crystal displays.   The characteristics of a single element are shown in Fig.4.

Notice that in contrast to devices produced in other laboratories, these devices pass on-currents greater than 1µA for source-drain voltages of 2V and gate voltages of the order of 10V.   Since it is important that any new device should be compatible with the existing voltages used in integrated circuits, this is not an unimportant point.   The off-currents in these devices [37] are typically 10^{-11} to 10^{-12} A so that the discharge of the liquid crystal elements will be determined by the liquid crystal display rather than the igfet.   Finally, the dynamic characteristics of these devices suggest that they are already sufficiently developed for application in 500 x 500 arrays [37].   Integrated circuit techniques have been used to fabricate a 7x5 array that successfully switches a liquid crystal display and further work is in progress in our laboratory and at RSRE to scale these devices to larger arrays.

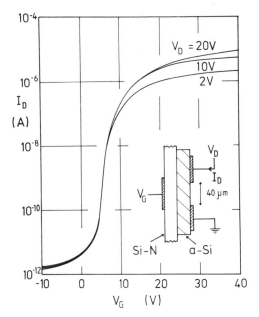

Fig.4 Drain current I_D plotted logarithmically against gate voltage V_G for the given values of drain voltage V_D. (after Ref.37)

References

1 P.G. Le Comber, W.E. Spear: Topics in Appl. Phys. **36**, 251–285 (1979)

2 W.E. Spear, P.G. Le Comber: Solid State Commun. **17**, 1193–1196 (1975)

3 W.E. Spear, P.G. Le Comber: Phil. Mag. **33**, 935–949 (1976)

4 P.G. Le Comber, A. Madan, W.E. Spear: J. Non–Cryst. Solids, **11**, 219–234 (1972)

5 A. Madan, S.R. Ovshinsky, E. Benn: Phil.Mag. **40**, 259 (1979)

6 A. Matsuda, S. Yamasaki, K. Nakagawa, H. Okushi, K. Tanaka, S. Iizima, M. Matsumura, H. Yamamoto: Jap. J. Appl. Phys. **19**, L305–L308 (1980)

7 R. Tsu, M. Izu, V. Cannella, S.R. Ovshinsky, F.H. Pollak: Proceedings 15th International Conference on the Physics of Semiconductors, Kyoto 1980.

8 A.J. Lewis, G.A.N. Connell, W. Paul, J.R. Pawlik, R.J. Temkin: Tetrahedrally Bonded Amorphous Semiconductors, ed. by M. Brodsky, S. Kirkpatrick, D. Weaire (American Institute of Physics, New York 1974) pp.27–32

9 A.J. Lewis: Phys. Rev. B. **14**, 658 (1976)

10 J.R. Pawlik, W. Paul: Proceedings of the 7th International Conference on Amorphous and Liquid Semiconductors, ed. W.E. Spear, (CICL, University of Edinburgh 1977) pp.437-441

11 W. Paul, A.J. Lewis, G.A.N. Connell, T.D. Moustakas: Solid State Commun. 20, 969-972 (1976)

12 G. Müller, S. Kalbitzer, W.E. Spear, P.G. Le Comber: Proceedings of the 7th International Conference on Amorphous and Liquid Semiconductors, ed. by W.E. Spear, (CICL Univ. of Edinburgh 1977) pp.442-446

13 P.G. Le Comber, W.E. Spear, G. Müller, S. Kalbitzer: J. Non-Cryst. Solids, 35 and 36, 327-332 (1980)

14 S. Kalbitzer, G. Müller, P.G. Le Comber, W.E. Spear: Phil. Mag. B. 41, 439-456 (1980)

15 W.E. Spear, P.G. Le Comber, S. Kalbitzer, G. Müller: Phil. Mag. B. 39, 159-165 (1979)

16 W. Beyer, R. Fischer: Appl. Phys. Lett. 31, 850-852 (1977)

17 M. Taniguchi, M. Hirose, Y. Osaka: J. Crystal Growth, 45, 126 (1978)

18 N. Sol, D. Kaplan, D. Dieumegard, D. Dubreuil: J. Non-Cryst. Solids, 35 and 36, 291 (1980)

19 W.E. Spear, P.G. Le Comber: J. Non-Cryst. Solids, 8-10, 727-738 (1972)

20 A. Madan, P.G. Le Comber, W.E. Spear: J. Non-Cryst. Solids, 20, 239-257 (1976)

21 A. Madan, P.G. Le Comber: Proceedings of the 7th International Conference on Amorphous and Liquid Semiconductors, ed. by W.E. Spear (CICL, Univ. of Edinburgh 1977) pp.377-381

22 N.B. Goodman, H. Fritzsche: Phil.Mag. B. 42, 149 (1980)

23 M.J. Powell: Phil. Mag. (in press)

24 M. Grünewald, P. Thomas, D. Würtz: Physica Status Solidi (b), 100, K139-143 (1980)

25 P. Thomas: private communication (1980)

26 R.J. Loveland, W.E. Spear, A.Al-Sharbaty: J. Non-Cryst.Solids, 13, 55-68 (1973/4)

27 D. Allan: Philos. Mag. B. 38, 381-392 (1978)

28 W.E. Spear, P.G. Le Comber, A.J. Snell: Phil. Mag. 38, 303-317 (1978)

29 D.A. Anderson, W.E. Spear: Philos. Mag. 36, 695-712 (1977)

30 P.G. Le Comber, W.E. Spear: Phys. Rev. Lett. 25, 509-511 (1970)

31 A. Gaith, P.G. Le Comber, W.E. Spear: (to be published)

32 D.I. Jones, R.A. Gibson, P.G. Le Comber, W.E. Spear: Solar Energy Materials, $\underline{2}$, 93-106 (1979)

33 See, for example, R. Fischer: Chp. 6 in "Amorphous Semiconductors" ed. by M.H. Brodsky, Topics in Appl. Physics, $\underline{36}$, 159-187 (1979)

34 See, for example, H. Fritzsche, Amorphous and Liquid Semiconductors, ed. by W.E. Spear, (CICL, Univ. of Edinburgh 1977) p.3

35 W. Beyer, R. Fischer, H. Wagner: J. Electronic Mater. $\underline{8}$, 127 (1979)

36 P.G. Le Comber, W.E. Spear, A. Ghaith: Electronics Letters, $\underline{15}$, 179-181 (1979)

37 A.J. Snell, K.D.MacKenzie, W.E. Spear, P.G. Le Comber, A.J. Hughes: To be published. A paper on this device development has been presented at the 10th European Solid State Device Research Conference, York, Sept. 1980.

The Effect of Hydrogen and Other Additives on the Electronic Properties of Amorphous Silicon

M.H. Brodsky

IBM Thomas J. Watson Research Center
Yorktown Heights, NY 10598 USA

1. Introduction

This tutorial discussion is a personal view on three types of additives that are of interest to the study of the electronic properties of amorphous silicon (a-Si). These additives are: 1) dangling and other weak bonds, 2) hydrogen and other bond terminators, and 3) n- and p-type dopants. Although dangling bonds are not generally thought of as additives, conceptually they can be viewed as extrinsic to a perfect continuous random network and therefore, as something added to that as yet unobtainable ideal. Either accidentally or by design, hydrogen, oxygen and more recently fluorine, have been used as bond terminators to passivate the extrinsic effects of dangling bonds. Quantitative analysis of H in electronically interesting amorphous silicon, namely a-Si:H, gives a typical concentration of order 10-20 atomic percent, a level indicative of a role for H that involves more than defect passivation. Substitutional dopants that are widely used in crystal silicon (x-Si), e.g. P and B, are now commonly added to a-Si:H to make n and p-type amorphous semiconductors. The doped materials are employed in diodes, transistors and other configurations of device potential. The assumption is that at least the electrically active P and B are substitutionally incorporated into the host Si network and therefore donate and accept electrons in ways analogous to shallow donors and acceptors in x-Si.

The discussion begins in Section 2 with microscopic models for the environment of dangling and weak bonds in pure a-Si:H and how H might then passivate the defects by shifting band gap states to other energy levels. In Section 3 a comparison between the properties of pure and hydrogenated a-Si is given and aspects of a-Si:H that are not easily explained by passivation alone are pointed out. In Section 4, which is the largest and most speculative part of the discussion, arguments for a quantum well model of the active role of H in a-Si:H are presented and the model is used to interpret effects beyond passivation. In Section V some inconsistencies in the picture of substitutional dopants are enumerated. Finally, Section VI contains concluding remarks about the present level of understanding of additives in a-Si.

56

2. Microscopic View of Dangling and Weak Bonds

Paramagnetic centers are observed in concentrations from 10^{18} to 10^{20} spins/cm^3 in nominally "pure" a-Si. The higher concentrations occur in sputtered, evaporated or ion bombarded layers that are prepared below 200 to 300°C. For cleanly evaporated a-Si deposited and measured at room temperature in ultra high vacuum [1], $N_s = 5 \times 10^{19}$ spins/cm^3. Higher temperature deposits, most notably chemical vapor depositions (CVD) by the pyrolysis of silane near 600°C, have spin densities at the lower end of the observed range [2], but these deposits are not documented as free from possible contaminants. The paramagnetic centers, often referred to as dangling bonds, may be associated with unpaired spins within vacancies or multi-vacancies or they may be associated with isolated unsaturated bonds distinct from multi-vacancy complexes [3]. However, it is known [4] that spin signals similar to and spanning the observed widths of the characteristic $g = 2.0055$ a-Si signal arise from vacancies and the like in x-Si. Although ball and stick networks with isolated dangling bonds can be built, there is no conclusive experimental evidence that favors (or for that matter, argues against) isolated dangling bonds over the more commonly accepted view of paramagnetic centers that are analogous to crystalline di-, tri-, and other multi-vacancies.

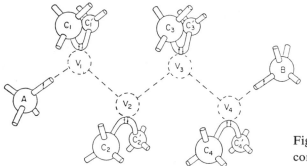

Fig.1 Tetravacancy in crystal silicon.

Figure 1 shows, for example, a crystalline tetravacancy with two dangling and four weak bonds. In crystal Si neutral multi-vacancies may have dangling bonds, as shown in Fig. 1, or may have their entire internal surface reconstructed into weak bonds with a net spin S=0. Those multi-vacancies which have no net spin when neutral, become paramagnetic when singly charged. For the point of illustrating the microscopics of hydrogenation, it is enough to consider the idealized planar representation of, for example, a tri-vacancy, as shown in Fig. 2. In the upper part of Fig. 2, the two aligned, but unpaired spins, would be observable as a single S=1 paramagnetic site. The S=0 state is more common for smaller vacancy complexes, S=1 is more commonly observed for larger complexes [5]. With hydrogenation, not only the dangling bonds, but the reconstructed weak bonds are likely to pick up H atoms as shown schematically in the lower half of Fig. 2. The adjacent back bonds (indicated by question marks) near the H terminated dangling bonds may also be weaker than a typical Si-Si bond within an ideal Si network (crystalline or amorphous) and therefore may also be easily hydrogenated. It is seen that many H atoms can be incorporated in the vicinity of a single paramagnetic defect. In fact, H can also be taken up by the non-paramagnetic multi-vacancies as well. We therefore can understand why a-Si:H can have of order 10-20 atomic percent H (i.e. $x = .1$ to $.2$ for formula $Si_{1-x}H_x$) while only of order 0.1% or less of the Si atoms seem to have paramagnetic dangling bonds in pure a-Si [6]. Furthermore, due to the nature of the disorder itself, there can be weak but diamagnetic bonds within a random network and these bonds may also be easily broken apart and then terminated by H atoms [7]. [Note: By a weak bond is meant a Si-Si bond with other than the ideal 2.35 Å interatomic separation of bulk solid Si.]

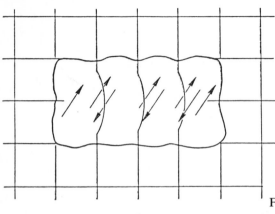

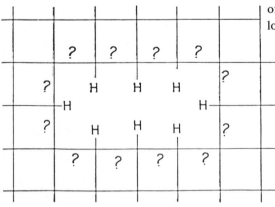

Fig.2   Idealized planar representation of a tri-vacancy: upper) In pure Si; lower) In hydrogenated Si.

It is well known for x-Si that electronic energy levels of the dangling and weak bonds of multi-vacancies and, in the limit of large multi-vacancies, of surfaces lie within the valence to conduction bandgap. Hydrogenation of x-Si surfaces removes gap states and replaces them with Si-H bonding levels deep down in the valence band and anti-bonding levels in or near the conduction band [8]. Thus, the first order effect of hydrogenation for a-Si is the removal of states from within the bandgap [9]. Many of the novel properties of a-Si:H follow as a consequence. The removal of states from the gap by hydrogenation is called passivation of the dangling bonds because the Si-H energy levels lie away from the bandgap and, therefore, presumably do not interfere with the near bandgap transport and optical phenomena of prime interest to semiconductor physicists and technologists. As will be seen in Section 3, the dangling bonds and related gap states obscure many such phenomena in pure a-Si. In Section 4 an additional role for H is proposed beyond that of passivation.

3. Comparison of Pure and Hydrogenated a-Si

Table 1 lists the comparative properties of interest for nominally pure and hydrogenated a-Si. By nominally pure is meant a-Si prepared by clean evaporation or by sputtering or ion bombardment with an inert species such as argon [10]. Except for the inert species, and perhaps some residual C or O at or below concentrations of oder 0.1 atomic percent, such preparations give pure a-Si. The column labeled hydrogenated refers to a product of the glow discharge decomposition of silane or the reactive sputtering of Si with a hydrogen bearing plasma. For either pure or hydrogenated a-Si we refer to typical substrate temperatures of 200 to 300C.

Table 1   A comparison of the properties of nominally pure and hydrogenated amorphous silicon.

	PURE	HYDROGENATED
Dangling Bonds	5×10^{19} cm^{-3}	$< 10^{16}$ cm^{-3}
States in the Gap	10^{19}-10^{20} cm^{-3}eV^{-1}	10^{16}-10^{17} cm^{-3}eV^{-1}
Conductivity	Hopping at Fermi Level $\exp - (T_0/T)^{1/4}$	Activated to Band or Band Tail $\exp - \Delta E/kT$
Optical Absorption	Tails into Infrared	Edge near 1.7 eV
Photoresponse	Nil	Photoconductive Photoluminescent
Doping Effects	Not Discernible	p- or n-type Conductivity Changes Up To 10^{10} Times

The first difference we note is that the ~ 5×10^{19} spins/cm^3 and the 10^{19} to 10^{20} gap states/cm^3/eV in pure a-Si are both reduced by orders of magnitude in a-Si:H

[4]. As a consequence the $T^{-1/4}$ Mott Law hopping conductivity attributable to carrier transport within the band gap is the dominant near room temperature conductivity mechanism in pure a-Si. For a-Si:H an activation energy, E_σ, characteristic of carrier activation to a band or band tail conduction path is observed [11]. Twice the intrinsic activation energy can be used to define a conductivity gap, i.e., $E_{cond} = 2E_\sigma^{max}$. For undoped a-Si:H, values of E_σ greater than ~ 1.0 eV have been observed [12], therefore $E_{cond} \gtrsim 2.0$ eV.

Because of the large number of gap states, pure a-Si is not characterized by a distinct optical absorption threshold but by an absorption tail that extends well below the $E_{xtal} = 1.1$ eV bandgap of x-Si. Furthermore, again presumably due to the gap states, the optical absorption coefficient α in the 1.1 to 2.0 eV region is orders of magnitude larger for pure a-Si than x-Si [10]. Explanations have been put forth to explain the excess absorption in terms of a disorder induced breakdown of the quasi-momentum selection rules that allow only phonon-assisted transitions to contribute to the indirect edge in x-Si [13,14]. At 1.5 eV, α about 10^3 cm^{-1} for x-Si, while an orders of magnitude higher α cm^{-1} would be expected for a 1.1 eV direct band gap semiconductor. For pure a-Si, α as high as 10^5 cm^{-1} has been observed at 1.5 eV. It should be noted that 0.9 eV potential fluctuations about 6 Å apart would be necessary to give such an enhancement from a disordered induced mechanism [14]. I believe that the dangling bonds and their associated gap states are the origin of this disorder rather than the milder potential fluctuations likely from the angular perturbations within an ideal, continuous random network. I shall return to this point later when an attempt is made to explain the energy at which the optical absorption rapidly rises in a-Si:H. The extrapolated threshold, ~ 1.7 eV, for the onset of strong absorption is often referred to as the optical gap E_{opt} of a-Si:H [11].

Two classes of phenomena that are not discernible for pure a-Si are easily observed for a-Si:H. These are 1) a sensitive photoresponse, i.e., low temperature luminescence [15] and room temperature photoconductivity [16] and photovoltaic [17] response, to visible radiation and 2) n- and p-type doping by trace impurities [18]. The study and potential applications of these phenomena is what makes a-Si:H so interesting. The low temperature photoluminescence spectrum peaks at $E_{pl} \approx 1.4$ eV with a full width at half height of about 0.5 eV. The room temperature photoconductivity follows the absorption edge.

To first approximation, the differences enumerated in Table 1 are explainable merely by the passivation role of H as a dangling bond terminator. Along with the removal of dangling bond states from the gap, the dominant conductivity and absorption processes are now associated with near band edge states or inter-bond transitions between them. Presumably because the in-gap states which acted as rapid non-radiative recombination centers in pure a-Si are no longer present after hydrogenation, observa-

60

ble lifetimes, even as long as milliseconds, are possible for photo-excited carriers that participate in, for example, photoconduction. Finally, with the reduction of the gap states, the addition of reasonable concentrations of dopants such as P or B can move the Fermi level and give either n- or p-type a-Si:H, respectively. However, the addition of 5×10^{19} H/cm^3, corresponding to the spin density in pure a-Si, is not the amount sufficient to give these interesting phenomena generally associated with a-Si:H. About two orders of magnitude more H is more typical [6]. What is all this extra H doing? We know that a good deal of it goes into the passivation of non-paramagnetic weak bonds, but even so, with so much H present the possibility of an active role beyond passivation becomes likely. One widespread view of the active role of H is as a homogeneous alloy constituent [12], that is, a-Si:H might be a $Si_{1-x}H_x$ alloy with a larger bandgap than pure Si. There is structural evidence [19,20] that the H is not homogeneously distributed. The next section discusses an alternative view based on the inhomogeneous potential fluctuations likely, even if the H in a-Si:H is homogeneously distributed. In particular, the observed ordering of measurable energies, namely, $E_{xtal} < E_{pl} < E_{opt} < E_{cond}$ is explained and some related predictions are made.

4. Quantum Well Model of a-Si With H

A recently proposed model [21] addresses the question of H in a-Si:H as a problem that involves inhomogeneities. The assumption is that the range of the potential disturbance about a Si-H bond in a matrix of pure Si is finite. Electrons, particularly those with near band gap energies, see distinctly different potentials near the Si-H bonds than they do near the Si-Si bonds well within a region of pure Si. The potential fluctuations are not averaged over to give a single bandgap, but rather different regions of space effectively have different band gaps.

The spatial scale of the fluctuations in potential, and therefore of bandgaps, is determined by the average distance between H sites or groups of H sites, say at a trivacancy. For a typical H concentration of 17 atomic percent, and about eight H atoms per average multi-vacancy, then the average distance between H-dressed cen rs is of order 10 Å. Within the context of percolation theory, extended state conduction (which is analogous to the existence of a delocalized state) can take place if the carriers can find paths around the Si-H potentials which here are assumed to be repulsive for both electrons and holes. Figure 3, which is based on an extended range percolation model for clusterings of H_2O molecules in supercooled water [22], gives a simplified two-dimensional picture of the effect of a finite range potential. In Fig. 3a, a square array of Si atoms is shown with a randomly chosen 5% of the Si-Si bonds broken and replaced by pairs of terminating H atoms.

61

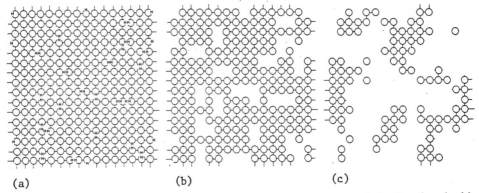

Fig.3   a) 2-D square lattice of Si atoms with two H atoms at each broken bond.   b) Those Si atoms with only Si nearest neighbors.   c) Those Si atoms without any nearest neighbor H atoms or any nearest neighbor Si atoms attached to H atoms.   This figure is based on a similar one for correlated sites of H_2O molecules in Ref. 22.

Now assume that because of the larger bonding-antibonding gap of Si-H bonds (compared to Si-Si) that carriers are excluded from the Si-H regions; for simplicity at first assume the range of an excluded region extends only as far as Si atoms that are part of Si-H bonds.   Figure 3(b) contains only the remaining Si atoms.   Although Fig. 3(a) appears to have fully interconnected Si regions, Fig. 3(b), which retains only those Si atoms connected to four other Si atoms, begins to show islands of pure a-Si with narrow interconnections.   Figure 3(c), where nearest neighbors to the Si atoms in Si-H bonds have also been removed, shows the islands more distinctly.   It is clear that for 17% H, a classical range of about 4 Å is enough to isolate Si islands.   By inspection of Fig.1, the islands would be about 2 to 9 atoms across, i.e., of order 10 ± 6 Å, at the percolation limit.   A qualitatively similar representation of the model could contain multi-vacancies rather than just broken bonds.   The model of Fig. 3 is conceptually different from structural clusters; nevertheless, when structural clusters of pure Si exist, they should exhibit behavior similar to the percolation defined regions of pure Si.   In 3-D, the percolation clusters are almost the same size as in 2-D.   In either case, the cluster size is more sensitive to variations in H concentration, possible superimposed structural inhomogeneities, and the assumed one atom range of the H potential disturbance than to dimensionality.   Only a rough estimate of the cluster size is made to illustrate the quantum effects that will become apparent below.   At this point what has been established is that there are regions of pure a-Si (clusters of Si atoms each in a Si environment) bounded and separated by potentials that emanate from Si-H bonds and that the cluster size is estimated to be ~ 10 Å from Fig. 3.   The cluster size and distinctiveness are model dependent; that is, they depend on the assumed potential range of the Si-H disturbance, the H concentration and its assumed clustering around multi-vacancies, the size of the multi-vacancies, and the type of simplified 2-D square lattice used for illustration.

In any direction a propagating electron will see a variety of pure Si regions with varying lengths of order 10 Å. Each Si regions will end with a barrier. The barrier height will be higher if the electron has to pass close to a Si-H bond and lower if the electron can leave a Si valley by going through a "pass" or saddle point between Si-H potential "mountains" or peaks. The net result will be a succession of wells and barriers of different lengths and heights, respectively. The situation is analagous to the ANDERSON-MOTT localization problem [23] and indeed at this point I resort to the accepted result that for each bond there is a mobility edge that is the same throughout the material. This result permits the representation of the varying barrier heights by a single barrier height for each band. COHEN [24] has proposed that 3-D quantum wells exist for near-band edge states in a binary alloy. Here only 1-D fluctuations are used to define quantum wells. More extended states can exist in the other two dimensions. It is to be understood that whenever a quantum level is refered to, what is meant is the lower bound of a band of states. Figure 4 then illustrates a typical pair of spatially coincident wells, one for electrons, one for holes. Note that the electron and hole wells coincide in space because they each are a consequence of the same set of barriers from the nearby Si-H bonds. In order to make quantitative estimates, the near gap band electronic structure within the pure a-Si clusters is approximated by that of x-Si. The barrier regions have a larger band gap and a valence band edge that lies significantly below the a-Si valence band. The conduction bands of the Si and barrier regions are closer to each other in energy. The schematic band structure of Fig. 4 follows. Along a direction through the a-Si clusters and the barriers the band gap is alternately E_x and $E_x + V_c + V_v$. Here V_c and V_v, where $V_c < V_v$, are the "effective" displacements in energy of the barrier conduction and valence band edges from the corresponding x-Si edges; V_c and V_v are each smaller than what would be the barrier heights for a head-on impact with a Si-H bond. The reduced barrier height, which is actually the mobility edge, is determined by how far away the Si-H bonds are from the island delimiting barrier and represents the minimum height of the saddle point between peaks centered at Si-H bonds. Any electron with energy less than V_c above the a-Si conduction band edge E_c will be confined to a potential well coincident with the a-Si region, any electron with greater energy will pass over the barrier.

Similarly any hole further than V_v below the a-Si valence band edge E_v will be delocalized, while holes with energies closer to E_v will be trapped. The quantum ingredient of the model is the "particle-in-a-box" nature of a trapped electron or hole, i.e., the bounded a-Si wells have quantized energy levels. Within an a-Si region the closest states to the gap will be ΔE_c above E_c and ΔE_v below E_v. The number and energy of the quantized levels depend on a well's dimensions; therefore, a macroscopic average gives a band of localized levels due to intra- and inter-well variations. For simplicity only one level, W_c or W_v deep, is shown in each band's well in Fig. 4(a). Schematical-

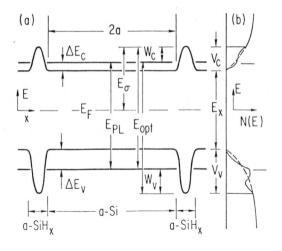

Fig.4   a) Real space band structure of an a-SiH$_x$-bounded region of pure a-Si. b) Schematic density of states N(E) vs energy E averaged over many regions.

ly, the spatially averaged density of states N(E) is shown in Fig. 4(b). The possible presence of a bump in N(E) depends on the actual distribution of well sizes in real a-Si:H.

In the density of states shown in Fig. 4(b), note that there are spatial correlations between the conduction and valence band localized state energies. States near the band edges are localized to larger ($>$ 10 Å) wells, while those states near the mobility edge are in wells with at least one small ($<$ 10 Å) dimension. For each conduction band state near the mobility edge, there's a valence band state, also near the valence band mobility edge, that is localized to the same quantum well. All the other localized states are also pair wise correlated in space and energy. Three such pairs are illustrated

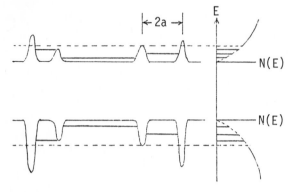

Fig.5   Schematic representation of the energies of the localized states in three quantum wells of different widths. The left side shows real space, the right shows the density of states with the pairs of localized states corresponding to each well. The dotted line is the effective barrier height, i.e., the mobility edge.

schematically in Fig. 5. The space and energy correlations are more general than the specific model here. They follow from the same arguments that lead to ANDERSON-MOTT localization. The same local potential fluctuation that leads to a valence band localized state will, if we assume the two bands have similar widths and overlap intervals, also lead to a conduction band localized state. Since the same local potential fluctuation is involved in both cases, then the two localized states must be near each other in space and furthermore, must have similar energies of localization as measured with respect to the mobility edge. While some of the effects of spatial correlation are implicit in the discussion below, there should be some additional consequences, particularly for interband transitions, that are calculable on the basis of the pair wise and higher order spatial correlations of localized states.

The logic of the argument so far has been that because of the range of the repulsion of carriers from the Si-H bond, the smallest effective barriers are V_c and V_v, but we know neither the range nor the barrier heights, exactly. We estimated that if the range of repulsion was of order one or two bond lengths, then the pure Si valleys are of order 10 Å. The other dimension of carrier confinement, namely the smallest energy of escape from the valleys will now be estimated from transport experiments.

From experiment [25], the mobility activation energies for electrons and holes, W_c and W_v, are about 0.2 and 0.35 eV, respectively. With a box size $2a \sim 10$Å and $m^*/m_e = 1$ and 0.5 for electrons and holes (from the heaviest x-Si effective masses), graphical solutions in 1-D let us use W_c and W_v to estimate $\Delta E_c \sim 0.1$ and $\Delta E_v \sim 0.25$ eV. Thus the estimated minimum separation between delocalized electrons and holes is $E_x + V_c + V_v \sim 2$ eV.

The N(E) of Fig. 4(b) is similar to that previously deduced [11]; to the extent that such a form for N(E) already works well for a-Si:H, so does the model. What is new is a microscopic model for an asymmetrical distribution of the localized states which are often called tail states and that the "tail" states (and, possibly, bumps) in N(E) are intrinsic, outside the band gap, and of the same order of magnitude as N(E) in x-Si. Presumably defect states also exist, but these lie in the gap and give orders of magnitude smaller contributions to N(E).

The model also gives many new explanations and predictions. As space permits some are enumerated below.

Optical Absorption: Using Fig. 4, I argue that the optical absorption is weak for transitions between the boxed-in states (localized to spatially coinciding V_c and V_v wells, but delocalized within each well) which retain some remnant of crystalline characteristics. By weak is meant of the same order of magnitude as in x-Si. An indirect bandgap is assumed to be present in the reciprocal space of the a-Si:H structure. Delocalized carriers, because they are of higher energy and therefore closer to each other in reciprocal space, are coupled to the localized states by the potential disturbances of the barriers; therefore, stronger, disorder induced absorption sets in when either one of the delocalized bands is involved [26]. Thus, $E_{opt} \sim 1.65-1.80 eV$ ($V_c + E_x + \Delta E_v$ or $\Delta E_c + E_x + V_v$). With doping there is less localization to the wells as evidenced by dopant related additional conduction paths (see below). Therefore, dopants lead to reduced ΔE's and therefore to absorption tails below E_{opt} which, in agreement with experiment [27], are more extended for p-type than n-type dopants not only because of the initially larger energy shift $\Delta E_v > \Delta E_c$ but also because the localized hole to extended electron transitions are the lowest energy contribution to E_{opt}. Furthermore, contrary to what would be the case from a Fermi level shift alone, the model predicts that doped, but compensated, a-Si:H should also show the more extended absorption tail.

Photoluminescence: When above band gap radiation excites electron-hole pairs, the excited carriers thermalize to localized states ΔE_c and ΔE_v away from what would otherwise be the band edges. For low temperatures, electron-hole pairs trapped within nearby spatially coincident conduction and valence band wells can radiatively recombine with a typical energy of $E_{pl} = E_x + \Delta E_c + \Delta E_v \sim 1.45$ eV. Note that $E_x < E_{pl} < E_{opt}$ as observed [11,15] for a-Si:H. A prediction is that, in general, amorphous or crystalline Si in appropriate barrier-isolated clusters of should give luminescence above E_x. Recent observations [28] on laser crystallized a-Si:H may be thus explained in terms of the exceptionally small x-Si particles formed.

Photoconductivity: The photoconductive response [16] of a-Si:H is activated with an energy comparable to W_c. The model is consistent with the thermally activated motion of photo-excited electrons from shallow wells, while the holes remain more deeply trapped. Photoconductivity below E_{opt} originates from the weak absorption tail in the same energy range as E_{pl}.

Electrical Transport: From the temperature dependence of electrical conductivity [12], activation energies E_σ^{max} higher that 1 eV have been observed for nominally undoped or compensated a-Si:H. The energy levels estimated for Fig. 4 imply that for a mid-gap Fermi level, electrons would require an energy $(E_x/2) + V_c \sim 0.85$ eV to be activated to extended states while holes would require $(E_x/2) + V_v \sim 1.15$ eV. Thus the maximum activation energy is $E_\sigma^{max} \sim 1$ eV and $E_{cond} \sim 2$ eV. As observed, $E_{cond} > E_{opt} > E_{pl} > E_x$. Corrections for the temperature dependence of the bandgap would have to be added to these arguments, for a closer quantitative comparison with experiment. However finer quantitative details are beyond the spirit of the presently used approximations.

The model is related to some of the existing band structure and localization concepts. Spatial fluctuations in band edges, either arising electrostatically or from

66

strains have been previously proposed [24,29]. In either case the possibility of E_x being retained as a <u>lower</u> limit on the band gap, with E_{pl} and E_{opt} then lying at higher energies because of quantum restrictions on the near gap levels, seems to have been missed. The closest analogy to the model is fluctuation induced ANDERSON-MOTT localization over an energy range near the top and bottom of a band. In fact, as pointed out above, classical percolation arguments are not rigorous enough to accurately predict localization, in essence the percolation model assumes ANDERSON-MOTT localized states exist. The extra physical ingredients here are: 1)the geometrical picture of how random fluctuations may enclose clusters, therefore there are spatial correlations between states at the bottom of the conduction band and top of the valence band, and 2)that, in contrast to band tailing models, for a-Si:H the fluctuations are all of the sign that expands the band gap, thus the near band edge localized states are further apart in energy than the unperturbed band gap.

5. Dopants

The conventional view [18] of doping in a semiconductor, and the view that has carried over as a first approximation to doped a-Si:H is that trivalent and pentavalent dopants (e.g., B and P, respectively) are incorporated substitutionally for Si with four-fold coordination. Therefore, B and P atoms are, respectively, acceptors and donors in a-Si:H as in x-Si. The host atomic structure and band structure are assumed to be negligibly changed by the presence of dopants. Within the host bandgap, the Fermi level E_f, is presumed to move according to how many electrical active donors (or acceptors) are present relative to the background density of states $N(E)$ at E_f. In a typical case, for low doping levels, electrical activity of about 1/4 the donors could account for the movement of E_f if $N(E_f)$ were of order the values deduced from field effect experiments, that is, $N(E_f) \sim 10^{17}$ states/eV/cm^3 for E_f near midgap. For the heavily doped case, impurity bands and new conduction paths are believed to occur [16,18].

There are several problems with the conventional assumption and their self consistency with experimental data. No resolution of the problems is proposed here; they are only pointed out as being important for study.

First there is only weak evidence [30] for four-fold coordination of dopants within the amorphous network. In fact, before SPEAR and LECOMBER [18] demonstrated p- and n-type doping in a-Si:H, it was widely accepted that B and P in ideal a-Si would not be forced into four-fold coordination because of the flexibility inherent in a random network structure [31]. The additional presence of H in a-Si could only serve to increase network flexibility and the known multivalences of B and P further reduce the likelihood of their four-fold coordination.

Ultraviolet spectroscopy of thin (~ 20 to 60 nm) films of doped a-Si:H has been used to deduce shifts in the average bandgap [32], that is the typical bonding to antibonding transition energy, as a function of doping. [The average bandgap as reflected in the UV absorption peak is to be distinguished from the threshold or minimum gap which is characterized by the absorption edge.] For of order 1% of P (n-type) or B (p-type) dopants, blue or red shifts, respectively, of about 0.1 eV are seen for the UV absorption peak (see Fig. 6). As discussed above in Section 4, the absorption edge also changes differently with n- or p-type doping. The edge and, in particular, the peak changes are much smaller or unobservable for x-Si. These spectral changes call into question the concept of an immutable host band structure for doped a-Si:H.

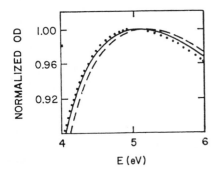

Fig.6 The normalized peak optical density of n-type (dashed line), nominally intrinsic (solid line) and p-type (dotted line) a-Si:H.

The quantitative arguments [18] for a large fraction (~ 1/4) of dopants being electrically active were based on the relatively large field effect density of states [11], $N(E_f) \sim 10^{17}$ states/cm^3/eV. It is now known [33] the surface and interface states make this an upper bound on $N(E_f)$ and values as small as $N(E_f) \sim 10^{15}$ states/cm^3/eV have been reported [34]. If the smaller value for $N(E_f)$ is right, then only a small fraction (< .01) of the dopants need be electrically active to move E_f through the observed range for light doping. The question then is what role do the large majority of electrically inactive dopants play.

It is important to note that the solubility limits for the common dopants in x-Si do not necessarily pertain to the amorphous phase. Presumably complete miscibility of B and As has been studied [35]. Therefore the intriguing possibility of a gradual transition from the doping to the alloying regime exists for a-Si:H. By the doping regime I mean a relatively fixed host (if that is possible) with donors and acceptors, while by the alloying regime I mean a continuously changing band structure that depends at least linearly on composition. From the evidence cited in this section, the alloying regime may be relevant even at the trace impurity level.

Finally, if the quantum well model holds true, then either or both the barriers and wells are likely to change with doping. The tailing of the optical absorption edge has already been discussed. No doubt transport and recombination processes also depend on how the dopants change the effects of inhomogeneities on the electronic structure.

6. Concluding Remarks

The theme of this presentation has been that the study of amorphous silicon also involves the study of additives. Even pure a-Si has the added ingredient of dangling bonds. Electronically important a-Si:H of course requires consideration of the role of H. The role of H extends beyond passivation of dangling bonds and other defects. In the future, useful interpretations should follow from the idea of spatially coincident quantum wells for holes and electrons. In particular, one should examine the dependence of the observable energies on H concentration and search for percolation threshold effects. Dopants also must play a role beyond the first approximation case of substitutional donors and acceptors. The intertwined effects of adding and subtracting, passivating and activating, doping and alloying are the basic ingredients of much of the current activity in the science and technology of amorphous silicon.

References

1. P.A. THOMAS, M.H. BRODSKY, D. KAPLAN, D. LEPINE: Phys. Rev. B **18**, 3059 (1978).

2. D. KAPLAN: In *Proc. Intern. Thin Film Congress* (Cannes, 1980) In Press. M. HIROSE, M. TANIGUCHI, T. NAKASHITA, Y. OSAKA, T. SUZUKI, S. HASEGAWA, T. SHIMIZU: J. Non-Cryst. Solids **35-36**, 297 (1980).

3. D. KAPLAN: *Inst. Phys. Conf. Ser.* **43**, 1129 (1979).

4. M.H. BRODSKY, R.S. TITLE: AIP Conf. Proc. **31**, 97 (1976); R.S. TITLE, M.H. BRODSKY, J.J. CUOMO: In *Amorphous and Liquid Semiconductors*, Ed. by W.E. SPEAR (University of Edinburgh, Edinburgh, 1977), p. 424.

5. Y.H. LEE, J.W. CORBETT: Phys. Rev. B **9**, 4351 (1974).

6. M.H. BRODSKY, M.A. FRISCH, J.F. ZIEGLER, W.A. LANFORD: Appl. Phys. Lett. **30**, 561 (1977).

7. M.H. BRODSKY, D. KAPLAN: J. Non-Cryst. Solids **32**, 431 (1979).

8. W.Y. CHING, D.J. LAM, C.C. LIN: Phys. Rev. Lett. **42**, 805 (1979).

9. B. VON ROEDERN, L. LEY, M. CARDONA: Phys. Rev. Lett. **39**, 1576 (1977).

10. M.H. BRODSKY, R.S. TITLE, K. WEISER, G.D. PETTIT: Phys. Rev. **B 1**, 2632 (1970).

11. W.E. SPEAR: In *Amorphous and Liquid Semiconductors*, ed. by J. STUKE and W. BRENIG (Taylor and Francis, London, 1975), p. 1.

12. T.D. MOUSTAKAS, D.A. ANDERSON, W. PAUL: Solid State Commun. **23**, 43 (1980); I. SOLOMON, J. PERRIN, B. BOURDON: Inst. Phys. Conf. Ser. **43**, 689 (1979).

13. J. TAUC: In *Amorphous and Liquid Semiconductors* ed. by J. Tauc (Plenum, New York, 1974), Ch. 4.

14. F. STERN: Phys. Rev. B **3**, 2630 (1971).

15. D. ENGEMANN, R. FISCHER: In *Amorphous and Liquid Semiconductors*, ed. by J. STUKE and W. BRENIG (Taylor and Francis, London, 1975), p. 947.

16. D.A. ANDERSON, W.E. SPEAR: Philos. Mag. **36**, 695 (1977).

17. D.E. CARLSON, C.K. WRONSKI: Appl. Phys. Lett. **28**, 671 (1976).

18. W.E. SPEAR, P.G. LECOMBER: Philos. Mag. **33**, 935 (1976).

19. J.C. KNIGHTS: J. Non-Cryst. Solids **35–36**, 159 (1980).

20. J.A. REIMER, R.W. VAUGHAN, J.C. KNIGHTS: Phys. Rev. Lett. **44**, 193 (1980).

21. M.H. BRODSKY: Solid State Commun. **36**, 55 (1980).

22. H.E. STANLEY: J. Phys. A. **12**, L329 (1979).

23. N.F. MOTT, E.A. DAVIS: In *Electronic Processes in Non-Crystalline Materials, Second Edition* (Oxford University Press, Oxford, 1979) Chapter 2.

24. M.H. COHEN: J. Non-Crystalline Solids **4**, 391 (1970).

25. P.G. LECOMBER, W.E. SPEAR: Phys. Rev. Lett. **25**, 509 (1970); A.R. MOORE: Appl. Phys. Lett. **31**, 762 (1977).

26. D. DIVINCENZO, M.H. BRODSKY: *To be published*.

27. J.C. KNIGHTS: AIP Conf. Proc. **31**, 296 (1976); E.C. FREEMAN, W. PAUL: Phys. Rev. B **20**, 716 (1979); M.H. BRODSKY, P.A. LEARY: J. Non-Cryst. Solids **35-36**, 487 (1980).

28. J.I. PANKOVE, C.P. WU, C.W. MAGEE, J.T. MC GINN: J. Electronic Materials **9**, 905 (1980).

29. H. FRITZSCHE: J. Non-Cryst. Solids **6**, 49 (1971); K.W. BOER: J. Non-Cryst. Solids **2**, 444 (1970); J. KREMPASKY, D. BARANCOK: In *Electronic Phenomena in Non-Crystalline Semiconductors* ed. by B.T. KOLOMIETS, (Academy of Sciences of the USSR, Leningrad, 1976), p. 185; S.M. RYVKIN, I.S. SHLIMAK: *Ibid.*, p. 203.

30. J.C. KNIGHTS, T.M. HAYES, J.C. MIKKELSON, JR.: Phys. Rev. Lett. **39**, 712 (1977); O.S. REILLY, W.E. SPEAR: Philos. Mag. B **38**, 295 (1978).

31. N.F. MOTT, E.A. DAVIS: *Ibid.*, First Edition (1971).

32. I. HALLER, M.H. BRODSKY: Inst. Phys. Conf. Ser. **43**, 1147 (1979).

33. N.B. GOODMAN, H. FRITZSCHE, H. OZAKI: J. Non-Cryst. Solids **35-36**, 599 (1980).

34. M. HIROSE, T. SUZUKI, G.H. DÖHLER: Appl. Phys. Lett. **34**, 234 (1979); J.D. COHEN, D.V. LANG, J.P. HARBISON: Phys. Rev. Lett. **45**, 197 (1980).

35. J.C. KNIGHTS: Philos. Mag. **34**, 663 (1976); B.G. BAGLEY, D.E. ASPNES, A.C. ADAMS, R.E. BENENSON: J. Non-Cryst. Solids **35-36**, 441 (1980).

New Insights on Amorphous Semiconductors from Studies of Hydrogenated a-Ge, a-Si, a-Si$_{1-x}$Ge$_x$ and a-GaAs

William Paul

Division of Applied Sciences, Harvard University
Cambridge, MA 02138, USA

1. Introduction

The presence of hydrogen during the process of deposition of thin films of amorphous tetrahedrally-coordinated semiconductors--whether an "accidental" component as in the glow-discharge decomposition of SiH$_4$ [1] or added deliberately to a sputtering plasma [2]--has become an accepted method of reducing the density of defect-related states in the pseudogap. It has been shown that the density of dangling bonds can be reduced by orders of magnitude and it is further supposed that the density of other types of defect (such as reconstructed weak bonds) may also be reduced. The success of the method has been clearly demonstrated for a-Si [1,2], a-Ge [2] and a-GaAs [3]. Nevertheless, there are quantitative differences between the results for a-Si:H and those for all of the other hydrogenated materials.

The gross effects of incorporation of H are clearly the same. They include (1) decreased electron spin density by several orders of magnitude (2) optical absorption and photoconductivity edges displaced to higher photon energies and often sharpened in slope (3) decreased electrical conductivity at all temperatures, by several orders of magnitude (4) activated electrical transport (5) increased magnitude of photoconductivity and photoluminescence. It does not seem to matter whether the samples are produced by sputtering in an argon-hydrogen plasma or by glow-discharge decomposition of gas mixtures.

A quantitative comparison of the results of hydrogenation of a-Ge and a-Si, presented in Table 1, strongly suggests that the efficacy of H in defect compensation is much greater for Si, despite the fact that deliberately hydrogenated sputtered a-Ge (and, in retrospect, a-Ge:H prepared by glow discharge decomposition of GeH$_4$) provided convincing evidence of the extent of elimination of the dangling bond defects in unhydrogenated evaporated or sputtered material [1,2].

Group 4-group 4 alloys are known to condense in continuous random networks, so that a-Si$_{1-x}$Ge$_x$:H may be regarded as the prototypical hydrogenated binary amorphous semiconductor. All such hydrogenated alloys display similar signatures of dangling bond elimination, but of degree much inferior to a-Si:H. Indeed, it appears that even slight alloying of C or Ge with a-Si:H reduces the effectiveness of hydrogenation. Yet such alloying is naturally thought of as a way of adjusting the energy gap for specific applications, such as solar cells, by analogy with crystalline semiconductors.

Table 1 Comparison of results of hydrogenation of a-Si and a-Ge for several properties. Some of the numbers are approximate. They are taken from various sources, but the differences for the two elements are nonetheless clear

	a-Si	a-Ge	Comment
Lowest reported spin density	10^{16}-10^{17}cm^{-3}	$>10^{18}$ cm^{-3}	Sputtered
Typical photoconductivity $\Delta\sigma$ at 10^{15} phot/cm^2-s	10^{-6}(1.96 eV) ohm^{-1}-cm^{-1}	5×10^{-8}(1.47 eV) ohm^{-1}-cm^{-1}	Sputtered 200-250°C
Best photoconductivity	2×10^{-5}(1.96 eV) ohm^{-1}-cm^{-1}	10^{-6}(1.47 eV) ohm^{-1}-cm^{-1}	Sputtered 250°C
Photoluminescence	Very efficient	Not reported	
Field effect D.O.S.	~10^{16}cm^{-3}eV^{-1}	$>10^{18}$cm^{-3}eV^{-1}	Glow discharge
$(\Delta G)_{RT}$ on hydrogenation	~× 10^{-8}	~× 10^{-4}	Sputtered
σ_0 in $\sigma=\sigma_0\exp[-E_\sigma/kT]$	10^{+3} - 10^{+4} ohm^{-1}cm^{-1}		Sputtered or Glow discharge
		300 ohm^{-1}cm^{-1}	Sputtered
		10^3 ohm^{-1}cm^{-1}	Glow discharge

Hydrogenated group 3-group 5 compounds, such as a-GaAs:H, also display unmistakable effects of dangling bond elimination, but no photoluminescence, very poor photoconductivity, and no field effect.

In order to try to obtain some insight into the question raised by the differences between a-Si:H and the other materials, we have prepared the hydrogenated elements and alloys of the Si-Ge system [4] and the GaAs system [5] either by sputtering in an argon-hydrogen plasma or by glow discharge decomposition, characterized them structurally and chemically, and measured their principal transport and optical properties. After surveying, in cursory fashion, the general character of the results, I shall focus on a few properties which seem particularly relevant for the establishment of a model to explain the observed poorer efficacy of H-incorporation in materials other than Si.

2. Preparation and Characterization

Our general *modus operandi* is to co-deposit many samples on a variety of substrates under controlled conditions, and after appropriate characterization, to attempt to correlate the measured properties among themselves and with the preparation conditions. Targets of 5" diameter, optical grade crystalline material are r.f. sputtered in an argon-hydrogen plasma. The base turbopump vacuum is between 3 and 5×10^{-7} Torr, with the residual gases established by an attached residual gas analyzer to be H_2O, CO, CO_2 and H_2. Both target and substrates are plasma-cleaned before deposition is begun. Several substrates of quartz, c-Si, Corning 7059 glass and Al are usually included for any one deposition. The substrate temperatures (nominal, by attached thermocouple) are between 200 and 400°C, the argon partial pressure is usually 5×10^{-3} Torr, and the partial pressure of H may be varied between 0 and 5×10^{-3} Torr. The flow rate of gases is controlled so as to maintain a constant ratio of gas pressures during a run, and the power is set so as to produce a deposition rate of about 1 Å/sec.

The amorphicity is checked by X-ray measurement on selected samples. Densities are checked, again on selected samples, by floating in a liquid of adjustable density. Thicknesses are determined by a Sloan Dektak profile-meter. The gross composition is established by electron microprobe and the profile *versus* depth in the sample of H, O, N, C etc. by secondary ion mass spectroscopy of some of the samples. Scanning and transmission electron microscope studies are carried out in a search for any heterogeneities. Finally, the H-content is estimated from analysis of the infrared vibrational absorption, cross-checked by studies of H-evolution on heating and nuclear reactions with ^{15}N on selected samples.

A large number of property measurements is usually carried out: infrared vibrational absorption between 200 and 3000 cm^{-1}, to establish the nature of the X-H stretching, bending and wagging vibrations; optical absorption edge spectra, to obtain a measure of the energy gap and the gap state density; electrical conductivity as a function of temperature; and steady-state photoconductivity under certain standard illumination conditions. In addition, when other data warrant, we study photoluminescence spectra, as a function of excitation photon energy, excitation intensity, temperature, and applied electric field; thermopower as a function of temperature; photoconductivity spectra and transient photoconductivity; field effect and capacitance-frequency measurements, to obtain an idea of the gap state densities; spin resonance and photo-spin-resonance; and finally, the evolution of H as a function of annealing temperature, which provides information on bonding and diffusion as well as on the H-content.

A simple dc glow discharge system was also used to prepare a-$Si_{1-x}Ge_x$:H with x between 0 and 1, from different ratio mixtures of SiH_4 and GeH_4 [4]. The total pressure of the combined gases was kept constant in the different preparations with the result that the rate of deposition at constant power varied between 0.1 µm/min and 0.5 µm/min as x was increased from 0 to 1. The compositional uniformity across the films was checked by electron microprobe to within the instrument sensitivity of 1%, and the actual composition determined to 2%. SIMS measurements showed uniform bulk densities of O less than 0.5% and of C less than 0.1%. These densities are somewhat higher than our usual finding for r.f. sputtered a-Si:H specimens. Sometimes, within 500 Å of the film--c-Si substrate interface of our glow-discharge produced samples, we found an increase of the O and C content. However, it is unlikely that any properties were substantially altered thereby, since Si-O infrared vibrational absorption was not observed even in 12 µm thick films. TEM and SEM micrographs showed smooth homogeneous regions with no evidence of columnar growth or other heterogeneity, at least down to a scale of 100 Å.

3. General Properties of a-Si$_{1-x}$Ge$_x$:H

For our glow-discharge preparations of these alloys, we find that many properties change monotonically, but nonlinearly, with x. Figure 1 shows a
selection of optical absorption edge spectra and Fig.2 the deduced variation
with x of (1) E_{04} and E_{05}, the photon energies at which the absorption
coefficient reaches 10^4 and 10^5 cm^{-1} respectively, and (2) E_g, the optical
energy gap found by extrapolation of $(\alpha h\nu)^{\frac{1}{2}}$ *versus* $h\nu$ to $\alpha = 0$, where α
is the absorption coefficient and $h\nu$ the photon energy. No corrections
for possible alterations with x of the H-content have been applied, since
edge spectra at fixed x and variable, known H-content are not available.
These results are different from those of CHEVALLIER, WIEDER, ONTON and
GUARNIERI [6], who found a linear variation, again without applying any
corrections for variation of H-content. As an aside, we note that ANDERSON
and SPEAR [7] and later SHIMADA, KATAYAMA and KOMATSUBARA [8], found that
the optical gap passed through a maximum with x in a-Si$_{1-x}$C$_x$:H. For this
alloy, however, there are possible complications due to 3-fold and 4-fold
coordinated C as well as possible variations in H-content.

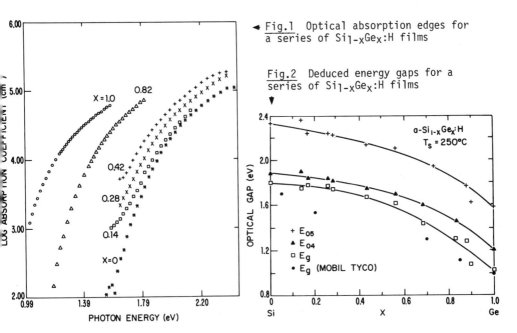

◄ Fig.1 Optical absorption edges for
a series of Si$_{1-x}$Ge$_x$:H films

Fig.2 Deduced energy gaps for a
series of Si$_{1-x}$Ge$_x$:H films

Figure 3 shows data on conductivity *versus* 1/T for several alloys. It
is seen that there is a wide range in T of activated transport, quite
adequate to define parameters in the equation for the conductivity σ:

$$\sigma = \sigma_0 \exp[-E_\sigma/kT] . \tag{1}$$

Figure 4 displays the deduced variation with composition of E_σ, σ_0 and
σ (room-temperature). The activation energy E_σ follows the same general
pattern as the optical energy gap: essentially unchanged near 0.85 eV for
x < 0.5 and then decreasing to 0.5 eV for x approaching 1. The values of

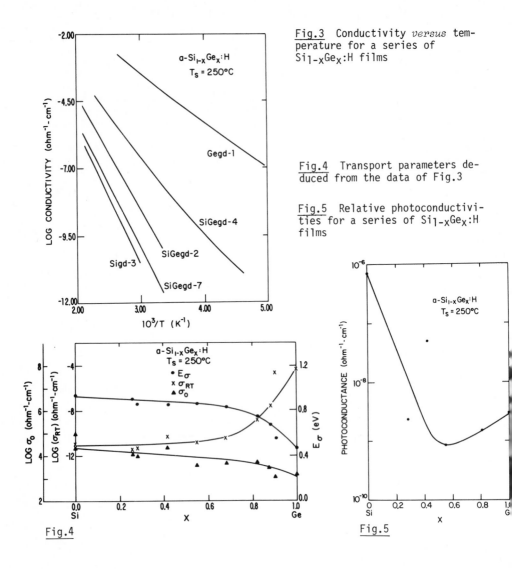

Fig.3 Conductivity *versus* temperature for a series of $Si_{1-x}Ge_x$:H films

Fig.4 Transport parameters deduced from the data of Fig.3

Fig.5 Relative photoconductivities for a series of $Si_{1-x}Ge_x$:H films

Fig.4

Fig.5

σ_0 are in the range commonly found for a-Si:H, and indicate that we are indeed studying transport in the conduction bands of these materials.

Preliminary data on photoconductivity and photoluminescence are shown in Figs.5 and 6. These show that the relevant magnitudes decrease very rapidly as x is increased from 0, which we believe agrees with the experience of other laboratories. The results again suggest, as we intimated in the Introduction, that the addition of Ge increases the gap density-of-states and deteriorates the photoelectronic properties.

There is very little, if anything, in the results to suggest to us *why* the photoelectronic properties are poorer in the alloys and a-Ge. In the

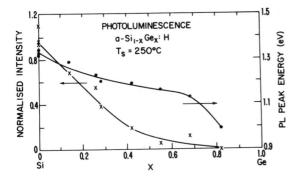

Fig.6 Relative photoluminescence intensities and peak photon energies for a series of $Si_{1-x}Ge_x$:H films

next section we shall describe the characterization of the H-content, H-complexes and H-bonding, as revealed by experiments on evolution of H on heating, infrared vibrational absorption spectra, and the changes in the infrared spectra at intermediate stages of anneal. We believe that these experiments *do* suggest the explanation we seek.

4. Selected Properties of $a-Si_{1-x}Ge_x$:H

(a) Evolution Spectra

Evolution spectra are taken on films that have been deposited on Al substrates and the Al subsequently dissolved away. An accurately weighed mass of amorphous sample is enclosed in a quartz container of accurately known volume. It is heated so that the temperature increases linearly with time. As gas evolves from the sample, the pressure in the enclosure increases.

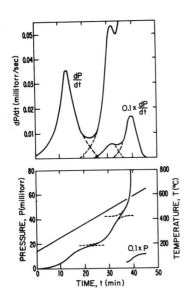

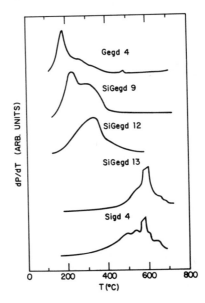

Fig.7 Typical H-evolution spectra *versus* time or temperature for the Harvard apparatus

Fig.8 Evolution of H from a series of $Si_{1-x}Ge_x$:H films

This is accurately measured with a capacitive manometer. The maximum pressure rise is to about 10^{-1} Torr, which is easily measurable. Two kinds of information are produced. The first, conveniently displayed as dP/dt *versus* T, tells us at what temperature the H is evolved from the sample. See Fig.7. The second is obtained from the total pressure rise. Knowing this, the volume of the system, and the mass of sample, the atomic percentage of hydrogen in the sample can be deduced [9].

A selection of evolution spectra is shown in Fig.8. We see that, whereas a large fraction of the H evolves from Ge near 150°C, the evolution from Si occurs above 400°C. The spectra for the alloys are intermediate but favor the spectra of the dominant component. Thus, there appears to exist weakly-bonded H in a-Ge:H and Ge-rich a-Si$_{1-x}$Ge$_x$:H alloys.

The concentration of H found from the total pressure produced is shown in Table 2. The alloys appear to contain more H than the end components.

Table 2 Characteristics of glow discharge a-Si$_{1-x}$Ge$_x$:H alloys

| Sample | Composition (Atomic %) | | | Preferential H Attachment, P |
| | Ge-content | H-content | | |
	Microprobe x	Evolution	IR Absorption Wag Mode	
Gegd-2	100	9	2.6	-
Gegd-4*	100	10	5.0	-
SiGegd-1	90	-	3.2	10.5
SiGegd-4	88	-	3.9	13.5
SiGegd-9	82	18	9.4	13.0
SiGegd-12*	70	38	23.0	9.7
SiGegd-8	68	10	9.1	11.6
SiGegd-14	65	13	9.5	10.3
SiGegd-3	55	-	8.0	12.5
SiGegd-2	42	-	11.4	10.4
SiGegd-6	28	-	10.3	8.2
SiGegd-7	26	-	12.0	14.0
SiGegd-13	14	15	9.2	~20
SiGd-4	0	8	10.7	-
SiGd-10*	0	15	11.6	-

*Samples deposited at $T_s = 100°C$; for all others $T_s = 250°C$

(b) Infrared Vibrational Absorption Spectra

The infrared spectra of the films were measured in the as-deposited condition and after successive anneals at several temperatures. Representative examples are shown in Fig.9.

The integrated areas under the different modes of vibrational absorption are informative. The integrated area under the wag mode near 600 cm^{-1} may be used to obtain the content of bonded H, by assuming equal matrix elements for absorption by Si-H and Ge-H wag modes, as suggested by the Stuttgart group [10]. These results are also displayed in Table 2; the H-content estimated from absorption is less than that found by evolution, which is itself (because of H diffusion into the quartz of the evolution chamber) a lower bound for the content. This again suggests the existence of weakly-bonded H, especially in a-Ge:H and a-Ge rich alloys.

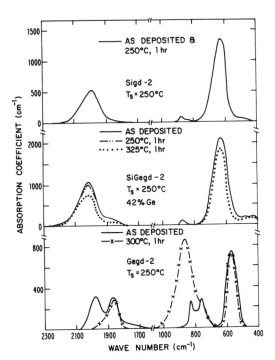

Fig.9 Infrared absorption for a series of $Si_{1-x}Ge_x$:H films, before and after successive anneals

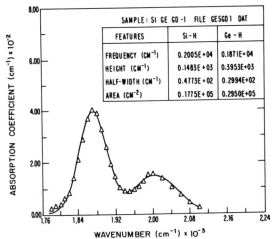

Fig.10 A typical computer fit (solid line) to the stretching vibrational optical absorption in a $Si_{1-x}Ge_x$:H alloy film

The graph labels for Fig.10:

SAMPLE: SI GE GD -1   FILE GESGD1 DAT

FEATURES	Si - H	Ge - H
FREQUENCY (CM^{-1})	0.2005E+04	0.1871E+04
HEIGHT (CM^{-1})	0.1483E+03	0.3953E+03
HALF-WIDTH (CM^{-1})	0.4773E+02	0.2994E+02
AREA (CM^{-2})	0.1775E+05	0.2950E+05

The absorption in the stretch modes near 2000 cm^{-1} may be deconvoluted to give the relative absorption by Si-H and Ge-H bonds, assuming the matrix elements are not changed by the local environments on alloying. A typical fit is shown in Fig.10. If we now evaluate the quantity:

$$P \equiv \frac{\text{integrated Si-H absorption}}{\text{integrated Ge-H absorption}} \div \frac{\text{Si content} \times \text{Si-H matrix element}}{\text{Ge content} \times \text{Ge-H matrix element}}$$

we get a measure of any preference P of H in bonding to Si rather than to Ge. The relative Si/Ge contents are given by the microprobe analysis. The assumption that the relative matrix elements for absorption by Si-H and Ge-H individual bonds [10], evaluated for the pure elements as 1.9, are not altered by changes in composition, seems a reasonable approximation in view of the near-equal electronegativities of the two elements. [As an aside, this assumption may be less reasonable in a-Si$_{1-x}$C$_x$:H alloys, as discussed by WIEDER, CARDONA and GUARNIERI [11].]

The preference ratio P, given in Table 2, is uniformly close to 10, a result that has great significance for the properties of these and other binaries, as we shall elaborate below.

Next we examine the changes in the spectra on annealing. For pure a-Si:H prepared at $T_S = 100°C$ or 250°C, no H is evolved by 250°C, and consistent with this, there is no change in the infrared spectra. The alloys and a-Ge:H are different. Thus annealing at 150°C for one hour of a-Ge:H prepared at 100°C evolves a considerable amount of H (see Fig.8), but there is no observed change in the infrared spectra. The same effect is present, although less marked, in the alloys. We conjecture that it is caused by the weakly-bonded H in a-Ge:H and alloy films, despite the fact that intuition suggests that H should usually induce some signature in infrared vibrational absorption. Our present glow discharge a-Si:H films do not show this effect, but we have reported, in 1979, that certain sputtering conditions appear to produce weakly-bonded H [12]. Identical conclusions have been reached, in parallel, by DENEUVILLE *et al.*, Grenoble [13] and more recently by ZELLAWA and GERMAIN in Paris [14].

Finally, annealing leads to the observation of Ge-O bands in our glow discharge a-Ge:H, but not in any others. This is taken to reflect a difference in the microstructure of the Ge films from the Si-alloy ones.

Before advancing a tentative explanation of these results, we shall review relevant experience with a-GaAs:H, and a-Si:H prepared under different sputtering conditions.

5. Selected Properties of a-GaAs:H Alloys

Our discussion of these alloys can be briefer, since we now wish to focus on certain properties suggested by the work on alloys of a-Si$_{1-x}$Ge$_x$:H. We have, in fact, carried out an even more extensive program of preparation, characterization and property measurement [3,5]. As before, hydrogenation leads to a displacement of the optical absorption edge to higher energies, the occurrence of activated transport, and an improvement in photoconductivity. These results are illustrated in Figs.11-13. There is no photoluminescence or field effect.

The infrared vibrational absorption spectra illustrated in Fig.14 are complex and relatively uninformative about specific modes. This is apparently a result of the occurrence of several Ga-H and As-H modes, occurring in a variety of local environments because of the Ga,As disorder.

The H-evolution spectrum is shown in Fig.15. It shows a very large low temperature peak. Moreover, an anneal which removes the H in this peak does not lead to proportionate changes in the infrared spectrum, suggesting for this material also the presence of weakly-bonded H.

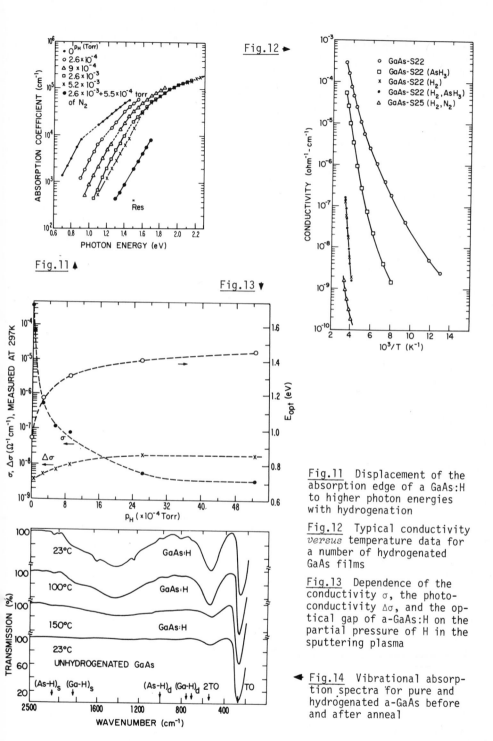

Fig.12 ➤

Fig.11 ▲

Fig.13 ▼

Fig.11 Displacement of the absorption edge of a GaAs:H to higher photon energies with hydrogenation

Fig.12 Typical conductivity *versus* temperature data for a number of hydrogenated GaAs films

Fig.13 Dependence of the conductivity σ, the photo-conductivity Δσ, and the optical gap of a-GaAs:H on the partial pressure of H in the sputtering plasma

◄ Fig.14 Vibrational absorption spectra for pure and hydrogenated a-GaAs before and after anneal

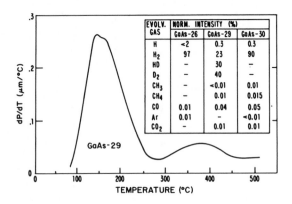

EVOLV.	NORM. INTENSITY (%)		
GAS	GaAs-26	GaAs-29	GaAs-30
H	<2	0.3	0.3
H_2	97	23	90
HD	–	30	--
D_2	–	40	–
CH_3	–	<0.01	0.01
CH_4	–	0.01	0.015
CO	0.01	0.04	0.05
Ar	0.01	–	<0.01
CO_2	–	0.01	0.01

Fig.15 Evolution of H from a-GaAs:H *versus* temperature

6. Selected Properties of a-Si:H Prepared at Different Partial Pressures of H

In our investigations of a-Si:H prepared by sputtering, we have made many samples in the (p_H, T_S) parameter space illustrated in Fig.16. For $T_S = 200°C$ and $p_H \leq 1$ mTorr, the H-content remains roughly constant at about 20 at.%.

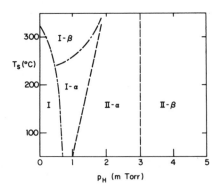

Fig.16 Parameter space for the production of sputtered a-Si:H alloys

However, there appear anomalies in practically all of the properties in Regions marked II-α and II-β; for reasons of space we shall concentrate on the transport results [15].

Figure 17 shows the transport results for Region I, where the H-content is increasing from zero: the conductivity changes regularly, monotonically and comprehensibly from non-activated, Fermi-level hopping for $p_H = 0$ to activated band transport for $p_H = 1$ mTorr. In contrast, the transport results for Region II-β are shown in Fig.18: although the H-content is the same as for $p_H = 1$ mTorr throughout this region, the conductivities are again high and resemble unhydrogenated material in their weak temperature dependence.

It is relevant at this point to describe some properties of samples prepared in Region II-β which concern our present theme. First, there is a relatively greater amount of bending modes at 890 and 840 cm^{-1}, and a relatively greater amount of 2090 cm^{-1} absorption than for samples in Region I.

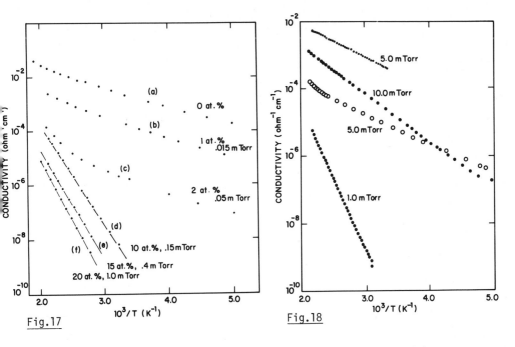

Fig.17 Conductivity *versus* temperature for a-Si:H samples prepared in Region I of Fig.16

Fig.18 Conductivity *versus* temperature for a-Si:H samples prepared in Region II-β of Fig.16

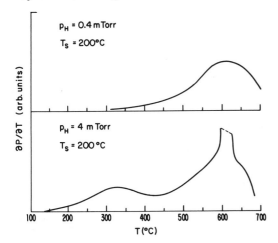

Fig.19 Evolution of H *versus* temperature for a-Si:H sample prepared in Regions I and II-β of Fig.16

Second, there is a tendency for the evolution spectrum to have more H in a low T evolution peak near 350°C, as is illustrated in Fig.19 [16]. Third, some fraction of the H evolved at this low temperature takes no part in infrared vibrational absorption, its evolution does not affect the optical absorption edge, and affects in a minor way the photoluminescence spectrum [17].

We have suggested a tentative interpretation of these results in terms of a heterogeneous structure for a-Si:H alloys [15]. The Si nucleates and grows in islands which eventually seek to join together, which leads to essentially a two-phase material of "islands" and connective "tissue" which are different in structure, chemical composition (H-content) and electronic band structure. This model originates in the generally-observed hetero-structure of films deposited from vapors, and the particular observations of microstructure by KNIGHTS and LUJAN [18] on Si-H films produced by glow discharge. In general terms, the transport must be explained on the basis of a percolation theory through two phases with quite different $\sigma(T)$.

7. Integrated Discussion of a-Si$_{1-x}$Ge$_x$:H Alloys, a-GaAs:H and a-Si:H of Region II-β

Because of (1) the similarities in the evolution spectra, which favor the release of H at rather low temperature, and (2) the occurrence of weakly-bonded H, as evidenced by the low T evolution and the absence of changes in the infrared spectra, we hypothesize that the structures of 4-4 alloys and a-GaAs resemble those of a-Si:H sputtered at high p_H [4]. In Si-H alloys we suppose that the H-attachment energy is such that Si-H bonds on the sur-faces of islands are preferred to reconstructed Si-Si bonds as the film grows. By contrast, we suppose that in a-Ge:H, the H-attachment is weaker so that reconstructed Ge-Ge bonds often win out on the island surfaces. Thus the H may be trapped in the connective tissue. The morphology of Ge, Ge rich alloys and GaAs is such that there is a greater defect density, concentrated in the connective tissue.

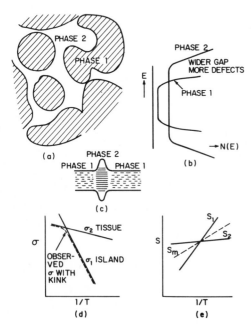

Fig.20 Hypothesized microstructure (a), density of electronic states N(E) *versus* energy (b), band struc-ture *versus* distance (c), conduc-tivity *versus* temperature (d), and thermopower *versus* temperature (e), for the islands (phase 1) and the H-rich connective tissue (phase 2) in a-Si:H films [15]

8. Summary

The more effective removal of pseudogap states in a-Si, by H compensation of defects, over other amorphous semiconductors may now be rationalized. It has several ingredients:

(1) the preference ratio for H-attachment to Si over Ge, a factor of 10, seems to be firmly established, given the argument that the ratio of matrix elements for absorption by Si-H and Ge-H bonds remains constant near the value determined by the Stuttgart group as the composition x of $Si_{1-x}Ge_x$ alloys changes. Such preferential attachment is not new in the amorphous semiconductor literature, since the same conclusion was reached earlier, by a somewhat different combination of experiments on $a-Si_{1-x}C_x$:H, by SHIMADA, KATAYAMA and KOMATSUBARA [8]. We believe it to be a general effect possible in all hydrogenated binary systems. It implies that defects connected with one element of a binary may be left uncompensated while H substitutes for the other element, which would provide adequate reason for the poorer photo-electronic properties of all binary alloys with respect to the pure elements. Preferential association of defects with the element unfavored for H-attachment should be detectable through electron spin resonance measurements on lightly hydrogenated binaries.

(2) preferential attachment in a binary does not explain the inferior compensation of defects by H in a-Ge itself. Our hypothesis, which is less firmly based than that concerning preference, nevertheless has the same genesis: it is that the Ge-H bond is less strong with respect to its competitors in a growing film (reconstructed, long Ge-Ge bonds, for example) than is the Si-H bond with respect to *its* competitors. The end result is a different microstructure which is more highly defected and contains relatively more weakly bonded H.

This model fits the experimental observations on H-evolution and the effect of annealing on infrared absorption spectra rather well.

(3) it is a corollary to our postulates of relatively weak Ge-H bonding that a different compensator might well make the properties of compensated Ge more similar to those for hydrogenated Si. We suggest the distinct possibility that O or F may be suitable. The incorporation of O has the additional advantage that it would promote the joining of growth islands and so reduce the volume of defected connective tissue. Improvement of binaries by this device might remain problematical if preferential attachment occurs for the new compensator; however it is possible that the binding to either element of a binary might still be stronger than the energy of alternative (weak) bonds in the network, so that the latter might still be reduced in density.

We are presently testing these ideas on Ge containing O and F, but have no results we wish to report at this juncture.

Acknowledgment

This work was supported in part by the U.S. Department of Energy under Contract DE-AC03-79-ET23036 and in part by the National Science Foundation under Grants DMR-78-10014 and DMR-77-24295.

References

1. R.C. Chittick, J.H. Alexander, H.F. Sterling: J. Electrochem. Soc. 116, 77 (1969); W.E. Spear, P.G. LeComber: Phil. Mag. 33, 935 (1976)
2. A.J. Lewis, G.A.N. Connell, W. Paul, J.R. Pawlik, R.J. Temkin: In A.I.P. Conference Proc., 20, 27 (1974); W. Paul, A.J. Lewis, G.A.N. Connell, T.D. Moustakas: Solid State Comm. 20, 969 (1976)
3. W. Paul, T.D. Moustakas, D.A. Anderson, E. Freeman: In Proc. 7th Int. Conf. on Amorphous and Liquid Semiconductors, ed. by W.E. Spear (CICL, University of Edinburgh, 1977), p. 334
4. D.K. Paul, B. von Roedern, S. Oguz, J. Blake, W. Paul: In Proc. 15th Int. Conf. on Semiconductors, Kyoto, 1980, to be published
5. D.K. Paul, J. Blake, S. Oguz, W. Paul: J. Non-Cryst. Solids 35 and 36, 501 (1980), and to be published
6. J. Chevallier, H. Wieder, A. Onton, C.R. Guarnieri: Solid State Comm. 24, 867 (1977)
7. D.A. Anderson, W.E. Spear: Phil. Mag. 35, 1 (1977)
8. T. Shimada, Y. Katayama, K.F. Komatsubara: J. App. Phys. 50, 5530 (1979)
9. S. Oguz and M.A. Paesler, to appear in Phys. Rev. (1980)
10. C.J. Fang, K.J. Gruntz, L. Ley, M. Cardona: J. Non-Cryst. Solids 35 and 36, 255 (1980)
11. H. Wieder, M. Cardona, C.R. Guarnieri: Phys. Stat. Solidi (b) 92, 99 (1979)
12. S. Oguz, R.W. Collins, M.A. Paesler, W. Paul: J. Non-Cryst. Solids 35 and 36, 231 (1980)
13. J.C. Bruyère, A. Deneuville, A. Mini, J. Fontenille, R. Danielou: J. App. Phys. 51, 2199 (1980)
14. Private communication
15. D.A. Anderson, W. Paul: to be published
16. S. Oguz: private communication
17. S. Oguz, J. Blake, R. Collins: private communication
18. J.C. Knights, R.A. Lujan: App. Phys. Lett. 35, 244 (1979)

Chemical Bonding of Alloy Atoms in Amorphous Silicon

Gerald Lucovsky

Department of Physics, North Carolina State University
Raleigh, NC 27650, USA

This paper presents a discussion of the local bonding arrangements of H and F in amorphous silicon. The basis for this model of the short range order derives from an interpretation of the local infrared absorption bands in a-Si:H and a-Si:F alloys that are associated with the incorporation of the respective alloy constituents, H and F. The method of analysis draws heavily on chemical bonding considerations and a review of the contributions to the bonding in Si-H and Si-F groups is presented. The approach to the interpretation of the ir-spectra is based on the use of induction relationships to derive realistic bond-stretching force constants for both Si-H and Si-F vibrations. These force constants change due to changes in near-neighbor bonding arrangements. This in turn allows one to interpret changes in bond-stretching frequencies in terms of changes in the local chemistry; i.e., the near neighbor environments. These bond-stretching force constants, along with bond-bending force constants obtained from various types of substituted silane molecules are used to calculate the frequencies and displacement vectors for localized Si-H and Si-F vibrations. In the case of Si-H, all of the modes of the SiH and SiH_2 groups are localized, and occur at frequencies above 500 cm^{-1}, the upper limit of Si network vibrations. For Si-F, the only modes that are localized are the bond-stretching vibrations which occur between 800 and 1025 cm^{-1}. The most important results presented in this paper are the assignments of the modes at 845 cm^{-1} in a-Si:H and 1015 cm^{-1} in a-Si:F. Both are associated with polymeric configurations of the general form $(SiX_2)_n$.

1. Introduction

It is well-established that the incorporation of either hydrogen [1] or fluorine [2] into amorphous silicon (hereafter a-Si) will reduce the number of electronically active defect states (traps) in the forbidden band gap and thereby enable p- or n-type doping via the incorporation of group III or group V elements, respectively [3]. This paper will address the way in which H and F-atoms are incorporated into an a-Si network, demonstrating how infrared absorption spectroscopy may be used to identify the various local bonding arrangements. Since the amount of bonded H or F necessary to eliminate defects, and thereby promote useful electronic properties is of the order of 2-20%, the systems that are dealing with are best described as binary alloys which we shall designate as a-Si:H and a-Si:F. The first section of this paper will deal with the question of chemical bonding, and the second and remaining portion with various aspects of the local vibrations associated with the alloy additives H and F.

There have been a number of detailed studies of the infrared absorption
and Raman scattering in a-Si:H alloys prepared by plasma decomposition of
SiH_4 [4-7], and by reactive sputtering of Si in a hydrogen containing ambi-
ent [4,8]. These studies have addressed two questions relating to the hydrogen incorporation as it is revealed through the observation of silicon-hydrogen (hereafter Si-H) local vibrational modes. The first is an association
of the various spectral features with different local bonding arrangements
such as SiH, SiH_2 and SiH_3 [4,7], and the second is a quantitative determination of the hydrogen concentration as computed from the strength of the
integrated infrared absorption. Other studies have considered changes in
the infrared absorption of the Si-H vibrations that are brought about through
the incorporation of additional alloy constituents such as oxygen [9], carbon
[10], boron [11] and most recently germanium [12], and by the annealing of
a-Si:H films at temperatures at which substantial hydrogen evolution is
known to occur [8,13]. This paper will emphasize the assignment of the
various spectral features to particular local bonding groups, and in addition point out some of the problems associated with the use of the integrated infrared absorption as a tool for the quantitative determination of
the hydrogen concentration.

Recently there has been an upsurge of interest in florinated a-Si
(a-Si:F), and in alloys that contain both hydrogen and fluorine, a-Si:H:F.
This interest stems from the observation that fluorine can play a role similar to hydrogen in eliminating mid-gap defect states, and thereby enabling
control of electronic properties via doping [14]. Research on the a-Si:F
and a-Si:F:H alloys has included numerous studies of the local vibrational
modes associated with the fluorine atoms [14-18]. The studies have emphasized the variation of the vibrational frequencies of bond-stretching modes
with the amount of incorporated fluorine. The various spectral features
have been in turn interpreted in terms of local bonding arrangements paralleling those previously used to interpret the spectra of a-Si:H; e.g.,
SiF, SiF_2 and SiF_3 groups.

This paper will summarize the infrared absorption data, and use a chemical
bonding approach for the assignment of the various Si-H and Si-F force constants. This represents an extension of a previous study in which variations in the Si-H bond-stretching frequencies were shown to be modified in
proportion to electronegatives of the atoms comprising the near-neighbor
environment to the particular Si-H group [19].

2. Chemical Bonding

We first compare the chemical bonding of H and the halogen atoms, F and Cl,
with the group IV elements, C, Si and Ge. It is convenient to discuss this
bonding in terms of three qualitatively different contributions which we
specify as covalency, bond-polarity, and π-bonding. It turns out that the
total bond energy can not be separated quantitatively into these three components in any simple way; never-the-less, there are qualitative differences
between Si-H and Si-F bonds which become apparent in such a comparison.

Table I provides a comparison of the properties of the bond energies for
C, Si and Ge with H, F. and Cl. For each of these bonds we give three numbers: (1) E_b, the bond energy in eV, [20], (2) ΔX, the electronegativity
difference using an electronegativity for each element that is an average of
the PAULING, SANDERSON and ALLRED-ROCHOW values [21], and (3) the so-called
π-bonding contraction, Δr, which is the difference between the bond-length

Table 1 Bond energy (e V), electronegativity difference, and π-bond contraction (Å) for bonds between C, Si and Ge and H, F and Cl

	H			F			Cl		
	E_b	ΔX	Δr	E_b	ΔX	Δr	E_b	ΔX	Δr
C	4.3e V ,	0.28,	0.0	5.0e V ,	1.40,	0.0	3.4e V ,	0.58,	0.0
Si	3.4e V ,	0.43,	0.0	6.0e V ,	2.20,	-0.15Å	4.0e V ,	1.29,	-o.03Å
Ge	3.0e V ,	0.12,	0.0	4.8e V ,	1.89,	-0.08Å	3.7e V ,	0.98,	-0.02Å

calculated from a set of empircal bonding radii and corrected for the electro-negativity difference, and the observed bond-length [20,21]. F or our pur-poses, the most important aspect of this table is derived from the comparison of Si-H and Si-F . Several things are obvious; first, the bond energy for Si-F is considerably greater than that of Si-H. This difference derives from at least two of the bond contributions identified above. The Si-F bond is significantly more polar than the Si-H bond. We will further quantify this difference by considering the partial charges on Si, F and H in the molecules SiF_4 and SiH_4. F ollowing the method of SANDERSON [22], and based on the uses of the stability-ratio (SR) electronegativities, the partial charges in SiF_4 are calculated to be e_{Si} = +0.68e, e_F = -0.17e compared to e_{Si} = 0.20e and e_H = -0.05 in SiH_4. F inally the large bond-length contrac-tion for Si-F is indicative of a π-bonding interaction involving back-dona-tion of electrons from 2p-states of fluorine (p_π) into otherwise unoccupied 3d-states of the silicon (d_π) [20,21]. This type of π-bonding interaction is only possible for four of the nine combinations identified in the table, Si-F , Si-Cl, Ge-F and Ge-Cl. F rom the values of Δr it is clear that this type of bonding contribution is strongest in Si-F . The magnitude of the π-bonding contribution will vary with the partial charge on the silicon, since the silicon 3d-orbitals necessary for the d_π p_π interaction have a spacial extent and energy that are both sensitive to the charge transfer or bond-polarity. This is exemplified by considering the variation of Δr with e_{Si} in the molecules $SiH_{4-n}F_n$ for n = 1, 2, 3, 4. F or example for SiF_4 e_{Si} = 0.68e and Δr = -0.15, whereas in SiH_2F_2, e_{Si} = 0.43e and Δr is decreased to -0.10. F or $SiHF_3$ the values of e_{Si} = 0.55 and Δr = -0.13 are intermedi-ate. The systematic changes in Si-F bond-lengths with the partial charge on the silicon atom also play an important role in explaining differences in Si-F bond-stretching frequencies (and force constants). F or Si-H bonds where π-bonding is not present, the Si-H bond-length also changes with the partial charge on the silicon atom via the bond-polarity. The surprising result is that the fractional changes in the vibrational frequencies for Si-F and Si-H vibrations with changes in e_{Si} are comparable even though the details of the bonding are qualitatively very different.

3. Local Atomic Arrangements

F igure 1 gives the local bonding arrangements that have been considered in the interpretation of the vibrational spectra of a-Si :H and a a-Si :F . Con-sider the first three types of bonding environments that have been labeled "isolated groups". The normal bonding coordination of Si is four, and that

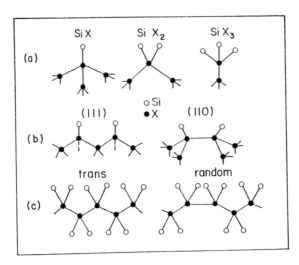

Fig. 1 Local bonding geometries: (a) isolated groups, (b) surface configurations, and (c) polymers

of H and F is one: therefore there are three ways in which H or F can be attached to a silicon atom which in turn is part of an infinite covalent network comprised only of other Si-atoms. These are, as SiX, SiX_2 and SiX_3 wherein X = H or F. These groups are designated respectively as monohydride (or monofluoride), dihydride (or difluoride) and trihydride (or trifluoride). The second portion of the diagram shows two "surface bonding" configurations, one having a (111) symmetry in which monohydride or monofluoride groups are attached to a second neighbor Si-atoms, and the second having a (110) symmetry in which monohydride or monofluoride groups are attached to nearest neighbor Si-atoms. The final portions of the figure indicate two different types of polymeric configurations based on the unit SiX_2. The first of these is a highly ordered polymer in which the hydrogen or fluorine atoms are in the so-called trans configuration. The second is a polymer configuration with random cis and trans local order. We will later demonstrate that the two most important bonding geometries as deduced from ir-studies are the Si-X group and the $(SiX_2)_n$ polymer.

4. Experimental Results and Interpretation, a-Si :H

This section summarizes the experimental results reported to date for a-Si:H alloys, and then discusses a methodology based on a force constant analysis for their interpretation in terms of the local environments discussed above.

In section 5 we proceed with a similar discussion for a-Si:F (and a-Si:F :H). Figure 2 gives a schematic representation of the ir-absorption spectra of a-Si:H alloys. It is compiled from the data presented in references [4-8]. The particular features which contribute to the spectrum of a given sample depend on the amount of incorporated Hydrogen as well as other deposition parameters [4-8]. A detailed discussion of this aspect of the ir-absorption studies is beyond the scope of this paper; however, it should be noted that very few samples, if any, contain only a single type of local environment with its unique ir-signature. Generally samples with low hydrogen content

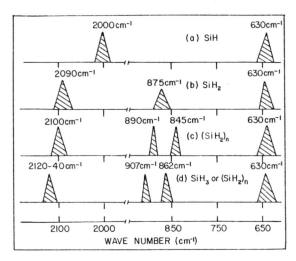

F ig.2 Schematic representation of experimental results, including proposed assignments [4-7]

(~3-5%) and produced on substrates held at temperatures near 300°C are domi-nated by absorption bands near 2000 cm^{-1} and 630 cm^{-1}, while samples contain-ing much more hydrogen, (~15-25%) contain additional spectral features near 850-900 cm^{-1}. These have been interpreted as being a signature of more than one H-atom being bonded to a given Si-atom, as for example in SiH$_2$ and SiH$_3$ groups. Also included in the figure are the assignments in terms of the various local bonding groups, SiH, SiH$_2$, (SiH$_2$)$_n$ and SiH$_3$. There is a con-census that an SiH group will give rise to two local vibrational modes which have frequencies that lie above the network vibrations of the a-Si host. These host vibrations extend to frequencies up to about 500 cm^{-1} [23] with distinct broad and overlapping peaks in three frequency regimes, 150-180 cm^{-1}, 280 cm^{-1}, 460-480 cm^{-1}. The localized SiH vibrations occur as a bond stretch-ing vibration at 2000 cm^{-1} and a bond-bending (or wagging) vibration at about 630 cm^{-1}. Support for this assignment will be derived from a force constant calculation using a cluster calculation. The second portion (Fig. 2) shows three features associated with an "isolated" SiH$_2$ group, a bond-stretching mode at 2090 cm^{-1}, a scissors bending mode at 875 cm^{-1}, and a broad band of rocking and wagging frequencies centered at 630 cm^{-1}. The polymerization of (SiH$_2$)$_n$ has been associated with two changes in the spectrum relative to that of SiH$_2$, a small shift to higher frequencies of the bond-stretching vibration, from 2090 to about 2100 cm^{-1}, and the appearance of a distinct doublet in the bond-bending regime with components at 845 and 890 cm^{-1}. The force constant model presented in this paper confirms an assignment in which the 890 cm^{-1} component of the doublet is due to a scissors motion, and the 845 cm^{-1} component to a wagging motion. The displacement vectors for these various types of vibrational modes are shown in [5]. The final portion of Fig. 2 gives features in the vibrational spectra assigned to the SiH$_3$ group [4,6]. In this paper we argue that a more probable assignment for these features is the polymerized group (SiH$_2$)$_n$. In this newly proposed assign-ment, the difference between the spectra in the two lower sections of Fig. 2 is believed to result from the degree of steric regularity. For example the spectra with the doublet at 845-890 cm^{-1} may be indicative of a random polymer, whereas as that containing the higher frequency pair at 862-907 cm^{-1} may be due to a polymer with a higher degree of order, as for example in a trans-conformation.

We have already indicated that a-Si and a-Si:H alloys exhibit additional broad and quasi-continuous spectra for frequencies below about 500 cm^{-1}. The incorporation of H into amorphous Si produces changes in this low frequency continuum [24] as well as the local modes discussed above. These have been interpreted as disorder induced changes in the network spectra [24]. An alternative explanation that will be discussed later in this paper is based on in-band resonance modes involving both H and Si motion. This explanation is supported by the Cluster calculations given in this paper. A better approach to in-band resonance modes is based on the Cluster-Bethe-Lattice Method [25,26]. In order to develop a force constant model to explain the frequencies of the various spectral features, as shown for example in Fig.2, it is necessary to have a set of reliable force constants for the various types of Si-H displacements, bond-stretching, bond-bending, etc. These force constants can in part be derived from force constant analyses of various silane molecules, provided that the chemical effects from differing local environments are properly considered.

It is well known in the chemical literature [27,28], that the frequencies of SiH bond-stretching vibrations depend on the chemistry of the other three atoms bonded to the Si-atom. This type of bond-induction effect has also been proposed for the bond-stretching vibrations of SiH, SiH$_2$ and SiH$_3$ groups in a-SiH [7,19]. Fig. 3 includes an analysis of molecular data on the SiH bond-stretching frequencies in substituted silane molecules, HSiR$_1$R$_2$R$_3$ wherein R$_1$ can be an atom such as F, Cl, Br or I or an organic group CH$_3$, C$_2$H$_5$, C$_6$H$_5$, etc. The vibrational frequencies have been plotted as a function of the sum of the electronegativities, X$_j$, of the R$_m$ atoms or groups. For purposes of illustrating this effect, the stability ratio electronegativities of SANDERSON [22] have been employed. The straight line shown in the figure is a least squares fit to the molecular data based on an equation of the form

$$\nu = \nu_0 + a\Sigma X_j \tag{1}$$

where the constants ν_0 and a have been found to be, ν_0 = 1740.7 cm^{-1} and a = 34.7 cm^{-1}. The error bar in the figure represents the uncertainty in the linear regression analysis, ±13 cm^{-1}. The points in the figure are for the experimentally determined values of the SiH bond-stretching frequency

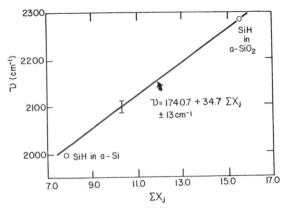

Fig. 3 Induction relation for SiH. The open circles are experimental data

in a-SiH, ν = 2000 cm^{-1} and ΣX_j = 7.86 and the SiH bond-stretching frequency in a-SiO$_2$, ν = 2280 cm^{-1} and ΣX_j = 15.63. In a network amorphous solid one expects the sum in (1) to have contributions from more distant neighbors as well as from the immediate neighbors to the SiH group. However, the results for the SiH bond-stretching frequencies in a-Si:H and a-SiO$_2$:H demonstrate that truncation of the infinite sum is possible, and that the molecular fit parameters, ν_0 and a can be used for amorphous network solids as well. The discussion presented in [19] demonstrates that the induction effect is a manifestation of decreases in the SiH bond-length that are induced by the charge withdrawal from the Si-atom as determined by the relative electro- negativity of its other three neighbors. The SiH bond-stretching frequency ν, and the SiH bond-length d are correlated by the relation

$$\nu d^3 = 7074 \ (cm^{-1}\mathring{A}^3) \tag{2}$$

For example in HSiF$_3$, ΣX_j = 17.25, ν = 2315 cm^{-1} and d = 1.44$\mathring{A}$, whereas in HSiSi$_3$ as in a-Si, ΣX_j = 7.86, ν = 2000 cm^{-1} and d is estimated to be 1.52$\mathring{A}$

The induction relationship explains the low value of 2000 cm^{-1} for the bond-stretching frequency of the SiH group in a-Si:H. In contrast, for sub- stituted silane molecules the corresponding SiH frequency ranges from 2100 cm^{-1} to a bout 2315 cm^{-1} [27,28]. In a-Si:H the three Si neighbors are less electronegative than any of the molecular groups or halogen atoms in the substituted silane molecules. Similar induction relationships have also been developed for the stretching frequencies of SiH$_2$ and SiH$_3$ groups [19]. At this time we emphasize the use of these induction calculations to obtain realistic force constants for valence force field calculations of the local mode frequencies.

There have been a number of different approaches used in calculating the vibrational frequencies of disordered solids. BELL and DEAN [29] have pio- neered calculations based on large clusters of the order of 500 atoms. Their most important contribution has been in the calculations of glasses with the local atomic arrangement of SiO$_2$. This method suffers from several problems: (1) it requires the diagonalization of very large matrices, 1500 x 1500 for 500 atoms; (2) it requires a procedure for establishing coordinates of the 500 atoms, i.e., building the network, and then demonstrating that it is properly relaxed, and finally (3) the vibrational density of states is sen- sitive to the boundary conditions imposed on the surface atoms. An alterna- tive approach which circumvents all of the problems sited above is based on the Cluster-Bethe-Lattice Method [25]. This approach removes the necessity for diagonalizing large matrices, building large clusters, and gives a way of generating realistic boundary conditions. The short-coming in this approach has been in generating an appropriate Calley-Tree and using realistic force fields. For the calculation of local mode frequencies it is possible to use yet a third approach based on clusters of intermediate size, $\sim$10 - 20 atoms. This procedure is viable because the local modes of interest involve dis- placements on only a small number of atoms, generally only those in the group of interest. This approach is particularly well-suited to the various Si-H groups, in a-Si since the atomic mass of hydrogen (m_H = 1.00) is very much smaller than that of Si (m_{Si} = 28.1). Surprisingly enough the model also works well for Si-F vibrations in a-Si:F alloys. The model calculation also admits the use of realistic force fields involving three-body non-central, as well as two-body central forces.

Figure 4 indicates two different types of clusters that we have used for calculating the properties of the bond-stretching and bond-bending modes of the SiH group. The 14-atom cluster is characterized by free atoms on the boundary of the cluster and no rings, whereas the 18-atom cluster has constrained atoms on the boundary and six-membered rings. The 18-atom cluster is generated from the 14-atom cluster by interposing a twofold-coordinated Si-atom between the three pair of the terminal atoms, and a threefold-coordinated Si-atom between the remaining three terminal atoms. These additional four atoms are fourth neighbors to the terminal hydrogen atom. In the model calculation we can increase the mass of these atoms from that of Si and simulate "fixed" type boundary conditions. Fig. 5 illustrates the type of force constants used in the cluster calculations. These are based on the valence force field method wherein the appropriate displacements involve either bond-lengths, Δr, or bond-angles, $\Delta \theta$. There is one type of two-body force k_r operative between each pair of atoms and three different three-body forces operative for each triad: a bond-bending force k_θ, a force involving simultaneous stretching of two neighboring bonds, $k_{rr'}$ and finally a force combining both stretching and bending motions, $k_{r\theta}$. We neglect $k_{r\theta}$ as being small with respect to the other three-body forces. Table 2 gives the values of the two and three-body forces we use in the calculations. The Si-H two body for k_r(Si-H) is calculated from the induction model whereas the two body Si-Si force k_r(Si-Si) is obtained from a fit to the lattice dynamics of crystalline Si [30]. The three-body forces involving triads of Si-atoms k_θ(Si-Si-Si) and $k_{rr'}$(Si-Si-Si) are obtained from the same analysis of phonons in crystalline Si. The bond-bending force k_θ(Si-Si-H) is obtained from a fit to the amorphous Si data, whereas, the remaining three-body force $k_{rr'}$(Si-Si-H) is obtained from a fit to molecular data. The cluster calculations imploying both 14 and 18-atoms yield identical results for the two local modes involving H-atom displacements. Table 3 summarizes these results, giving the calculated frequency and the atomic displacements of the H atom and its immediate Si-neighbors. The bond-stretching motion involves out-of-phase displacements of the H and Si atoms with the major displacement being for the H-atom. This is a simple reflection of the relative masses of the two atoms. The bond-bending mode involves H motion in a direction perpendicular to the Si-H bond. Conservation of momentum requires a small displace-

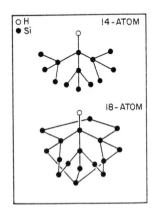

Fig.4 Local clusters for SiH calculations

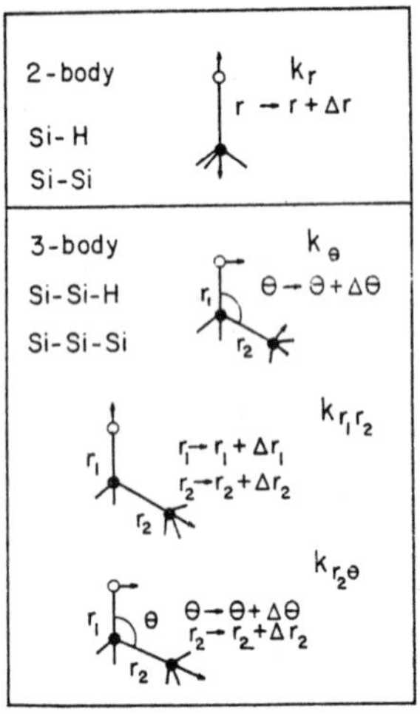

Fig.5 Valence forces for SiH calculations

Table 2 Force constants (10^5 dynes/cm) for a Si :H

2-Body	k_r(Si-H)	2.30
	k_r(Si-Si)	1.52
3-Body	k_θ(Si-Si-H)	0.091
	$k_{rr'}$(Si-Si-H)	0.030
	k_θ(Si-Si-Si)	0.092
	$k_{rr'}$(Si-Si-Si)	0.012

Table 3 SiH frequencies and displacements

Mode	Frequency (cm^{-1})	Displacements (Arbitrary Units)	
		H	Si
Local Modes			
Stretching	2000	0.97	-0.03
Bending	630	0.76	-0.16
Resonance Modes			
Shear	325	0.03	0.24
Stretch	265	0.02	0.42

ment on the second-neighbor Si-atoms as well. The calculations also yield in-band resonance modes wherein the Si and H displacements are in phase. These occur as shearing modes in frequency regime between 320 and 330 cm^{-1} and stretching modes in the frequency regime between 250 and 275 cm^{-1}. These modes are slightly different for the 14-atom and 18-atom clusters, and will be subjected to further study using the Cluster-Bethe-Lattice approach [26].

As we have pointed out above, there is also considerable interest in the vibrational modes of both isolated SiH$_2$ groups and SiH$_2$ groups in polymerized configurations, (SiH$_2$)$_n$. The vibrational modes of an isolated SiH$_2$ group are shown in [5]. For the relatively small mass ratio m_H/m_{Si} = 0.03, the frequencies of the symmetric and asymmetric bond-stretching modes are near degenerate. We have applied cluster calculations to isolated SiH$_2$ groups and to polymerized configurations as well. For the isolated groups we have used clusters paralleling those used for SiH, an 11-atom cluster with free atoms at the boundary and no rings, and a 20-atom cluster with constrained atoms on the boundary and six-membered rings. Fig.6 emphasizes the difference in the near-neighbor SiH$_2$ groups for isolated and polymerized configurations. These are important in fixing the various force constants.

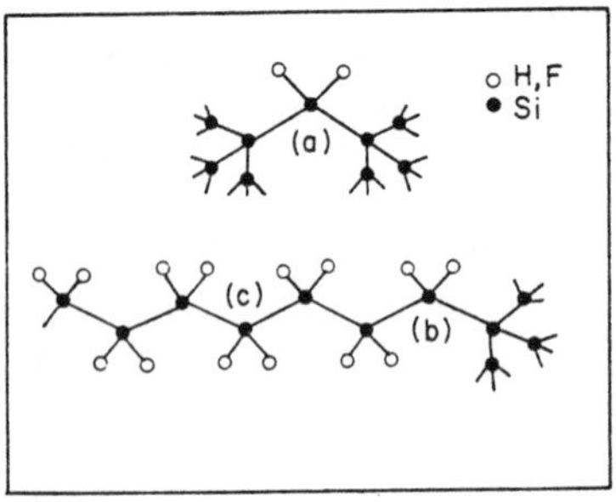

Fig.6 Local environ-
ments for SiH_2; SiF_2:
(a) isolated group,
(b) terminal group
in $(SiH_2)_n$, and
(c) mid-group in
$(SiH_2)_n$

For an isolated SiH_2 group the bond-stretching force constant is obtained via the induction calculation: $k_r(Si-H) = 2.5 \times 10^5$ dynes/cm. The bond-bending forces are the most important in discussing the vibrational proper-ties of SiH_2 groups. The bond-bending force constant involving the triad H-Si-H is obtained from molecular spectroscopy, i.e. $k_r(H-Si-H) = 0.185 \times 10^5$ dynes/cm. whereas the remaining three body bond-bending force $k_\theta(Si-Si-H)$ is taken to be equal in value to that used for the isolated SiH group, 0.09×10^5 dynes/cm. Table 4 summarizes the results of the calculations for the 11-atom and 20 atom clusters. These calculations support the assign-ment discussed earlier in this paper. Namely, that the 875 cm^{-1} mode is the scissors type bond-bending vibration, and that the 630 cm^{-1} mode contains contributions from both wagging and rocking motions. The twisting mode is ir-inactive. In these model calculations the only mode that varies in fre-quency between the two clusters is the bond-rocking mode which has a con-siderable contribution from motions of second-neighbor Si-atoms.

Table 4 Local vibration - isolated SiH_2

Mode	11-Atom Cluster (cm^{-1})	20-Atom Cluster (cm^{-1})
Stretching	2088	2088
Scissors	880	880
Wagging	651	654
Twisting	609	609
Rocking	501	517

Consider the next changes that occur in going from an isolated SiH_2 group to one in the central portion of a polymer segment $(SiH_2)_n$. The bond-stretching force constant is increased due to an effective increase in the electronegativity of the second-neighbor Si-atoms (see Fig.1). For the isolated SiH_2 groups the electronegativity sum in the analog of (1) is 5.24 corresponding to Si-atoms with all Si-neighbors, whereas for the polymer-centered group, the sum is increased to 6.41 reflecting the fact that each second-neighbor Si-atoms has two H-neighbors and one Si-neighbor. This has the effect of increasing the bond-stretching force constant from 2.50×10^5 dynes/cm to 2.57×10^5 dynes/cm. Comparisons with calculations for disilane Si_2H_6 and other silane molecules, indicate that $k_\theta(H-Si-H)$ is not strongly influenced by changes in the bonding to the central H-atom. We therefore take this force to be the same as in the isolated group, 0.185×10^5 dynes/cm. These same comparisons indicate substantive changes in the magnitude of the "other" bond-bending force $k_\theta(Si-Si-H)$. For the isolated SiH and SiH_2 groups this force is the same, 0.091×10^5 dynes/cm whereas for the disilane molecule it is increased by more than a factor of two, to 0.184×10^5 dynes/cm. The key factor seems to be the number of hydrogen neighbors attached to the second Si atom. A scaling argument based on this assumption yields a value for $k_\theta(Si-Si-H) = 0.15 \times 10^5$ dynes/cm for the SiH_2 group in the interior portion of a polymer segment, $(SiH_2)_n$.

Table 5 compares the results of the calculations for isolated and polymerized SiH_2 groups with the previously proposed assignment, Fig.2. The calculations are fully supportive. Also included in the table are the other assignments for SiH_3 or $(SiH_2)_n$. The comparisons in Table 5 suggest that these may indeed be due to $(SiH_2)_n$ rather than SiH_3.

Table 5 Local vibrations - SiH_2 and $(SiH_2)_n$

Mode	SiH_2 (cm^{-1})		$(SiH_2)_n$ (cm^{-1})	
	ν Calc.	ν Exp.	ν Calc.	ν Exp.
Stretch	2088	2090	2108	2100-2120
Scissors	880	875	910	890,910
Wagging	651	630	821	845,861
Rocking	501	-	538	630

In summary for a-Si:H we can go far in explaining the experimental data with three local groups, SiH, SiH_2 and $(SiH_2)_n$. However, care must be exercised in not carrying these assignments too far. For example some workers have referred to the 2100 cm^{-1} mode as being a dihydride vibration. The calculations summarized in Tables 4 and 5 support this assignment. It must be noted that an assignment for the 2100 cm^{-1} absorption as SiH_2 requires the simultaneous **occurrence** of absorption between 840 and 910 cm^{-1}, with one component being the scissors mode of the SiH_2 group. Modes can also occur near 2100 cm^{-1} but with no absorption near 850-900 cm^{-1} [31]. These have been emphasized in a recent paper by PAUL [31]. Modes of this sort can come about in at least two ways. One can have an SiH group in an environment wherein one of the Si-neighbors is a strongly electronegative atom, for example oxygen. The frequency of this SiH vibration is increased from 2000 cm^{-1} to 2090 cm^{-1} [8] and does not involve absorption near 850-900 cm^{-1}. Alternatively, it has been suggested in [31] that there may be other SiH configurations in which additional couplings may produce modes near 2100 cm^{-1}. This point needs additional study, both via experiment and model calculations.

So far we have emphasized the association of the observed spectral features with various local bonding geometries. A second aspect of ir-studies involves attempts to use the integrated absorption as a measure of the total bonded hydrogen content. This is wrought with many difficulties [32]. Partial success has been achieved using the integrated absorption in the bond-bending or wagging regime, at approximately 630 cm^{-1} [33]; however further work is necessary. A detailed discussion of some of the problems, for example, various aspects of spatial non-uniformities in the hydrogen incorporation, is beyond the scope of this presentation.

5. Experimental Results and Interpretation, a-Si:F

We use the same approach in this section as we have done above for a-Si:H. We review the experimental data and then proceed to provide an interpretation that is based on a combination of chemical bonding considerations, i.e. induction, and force constant calculations. Before embarking on these we first consider some qualitative aspects of the vibrational modes of the various SiF groups that result from the increased mass ratio of m_F/m_{Si} (= 0.68) compared to that of m_H/m_{Si} (= 0.03), and from the increased bonding energy, E_b(Si-F) = 6.0 eV as compared to E_b(Si-H) = 3.4 eV. The increased mass ratio means more Si motion in the local vibrational modes whereas the increased bond-strenth implies greater force constants. We also anticipate larger changes in bond-induction effects due to the π-bonding contribution that is present in Si-F bonds.

Consider first the effects of the increased mass ratio. For the bond stretching vibrations of SiH and SiH$_2$ groups, the ratio of H to Si displacements was large, of the order of 10:1 or more. The calculations made for SiF and SiF$_2$ groups on the other hand indicate nearly equal displacements for each atom. Whereas all of the modes of the various SiH groups shown in [5] are local modes and occur at frequencies greater than those of the host a-Si network, all of the vibrations of the various SiF groups are in-band resonance modes except for the bond-stretching vibrations. The stretching modes occur between 800 cm^{-1} and 1025^{-1}, while the bending, rocking and wagging modes are all below 400 cm^{-1}.

A second manifestation of a larger mass ratio is in a splitting of the symmetric and asymmetric modes of the SiF$_2$ and SiF$_3$ groups. For the SiX$_2$ groups, the ratio of the frequencies ν_a/ν_s (which is always greater than one) is approximated by the relation given below,

$$\left(\frac{\nu_a}{\nu_s}\right)^2 = \frac{(1 + 2\bar{m}\sin^2 2\alpha)(1-\bar{k})}{(1 + 2\bar{m}\cos^2 2\alpha)(1+\bar{k})} \tag{3}$$

where $\bar{m}$ is the mass ratio m_X/m_{Si}, $\bar{k}$ is the ratio of three-body to two-body stretching frequencies [k_{rr}'(X-Si-S)/k_r(Si-X)], and 2α is the angle between the two X-Si bonds, i.e. the X-Si-X bond angle, which is taken to be tetrahedral, 109.47°. Table 6 summarizes the results of this relation as application to SiH$_2$, SiF$_2$ and SiCl$_2$ groups. A comparison of the results of the calculation with splitting of the SiX$_2$ stretching frequencies in protypical substituted silane molecules supports the model from which (3) was derived. A detailed analysis of the factors contributing to the splitting, i.e., the terms in the brackets in (3), shows that the determinant factor in promoting the large increases in ν_a/ν_s for SiF$_2$ and SiCl$_2$ as compared to SiH$_2$ is the increased mass ratio, $\bar{m}$.

Table 6 Ratio of asymmetric to symmetric stretching frequencies for local SiX_2 groups

Group	$\overline{m}$	$\overline{k}$	Calculation (cm^{-1})	Experiment (cm^{-1})
SiH_2	0.035	0.01	1.003	$1.00, H_2Si\ Cl_2$
$Si\ F_2$	0.676	0.05	1.10	$1.09, F_2Si\ Br_2$
$SiCl_2$	1.264	0.05	1.14	$1.10, Cl_2Si\ H_2$

There have been a number of studies of Si-F vibrations in a-Si:F and a-Si:F:H alloys [14-18]. There is a consensus among these workers regarding several of the assignments, for example the SiF bond-stretching mode is assigned to the absorption at 830 cm^{-1}, whereas there is considerable disagreement about the nature of the relatively strong and sharp absorption at 1015 cm^{-1}. It has been attributed to an SiF_3 absorption by SHIMADA and his coworkers [16,18], and an SiF_4 molecule by MPI group [17]. We first argue against these assignments and then go on to present a systematic set of arguments that establishes a similar behavior between a-Si:F and a-Si:H; i.e., that the dominant contributions to the ir-absorption are from SiF at low F contractions, (<10%F), and $(SiF_2)_n$ at higher concentrations (>20%F).

Consider first the arguments against the SiF_3 model. In this the symmetric and asymmetric stretching modes of the SiF_3 group are assigned as 838 cm^{-1} and 1015 cm^{-1}, giving a ratio, $\nu_a/\nu_s = 1.21$. Using a relation similar to (3)

$$\left(\frac{\nu_a}{\nu_s}\right)^2 = \frac{(1 + 1.5\overline{m}\sin^2\beta)(1-\overline{k})}{(1 + 3\overline{m}\cos^2\beta)(1+2\overline{k})} \qquad (4)$$

where β is the angle between an Si-F bond and the threefold symmetry axis of the SiF_3 group. We calculate a ratio of 1·16 which compares to experimental values of 1·10 in $BrSiF_3$ and 1·13 in Si_2F_6. For Si_2F_6, we "average-out" the effects due to the molecular symmetry. If 838 cm^{-1} were the symmetric mode frequency, then the frequency of the asymmetric mode would be at most 970 cm^{-1}. The arguments for SiF_4 hinge on the occurrence of two modes at frequencies close to those of the SiF_4 molecule. A test for this model is the occurrence of a polarized Raman mode at 800 cm^{-1}. There has not been reported in any samples of a-SiF making this assignment unlikely.

Consider that the experimental data for local modes in a-Si:F and a-Si:F:H. Table 7 summarizes these data. There is a consensus that the mode at 830 cm^{-1} is the stretching mode of the SiF group, and a further consensus that those modes for which $\nu>900\ cm^{-1}$ are associated with local environments having more than one F. In this work we argue that these are all SiF_2 groups, either isolated or in polymeric configurations, $(SiF_2)_n$.

We first consider the assignment for the SiF bond-stretching vibration and use an induction argument to predict its frequency. Fig.7 shows the induction relation deduced from data on various substituted silane molecules containing SiF groups, $SiFH_3$, $SiFCl_3$, $SiFClBrI$, etc. The solid line represents the fit to the data, with the numerical result shown in the figure as

Table 7 Stretching modes in a-Si :F

Frequency (cm^{-1})	Assignment	References
828,850	Si F	[14, 15, 16, 17]
827,870	Si F $_2$, symmetric	[16]
920,965	Si F $_2$, asymmetric	[16]
838,1015	Si F $_3$	[16]
1015	Si F $_4$	[17]

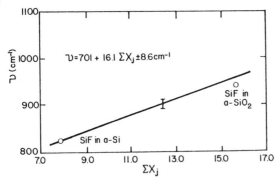

Fig.7 Induction relationship for SiF

well. The experimental points shown in the figure are for SiF in a-Si:F and SiF in a-SiO$_2$:F. They support the use of molecular data for the network solids, essentially the same thing we have previously noted for SiH vibrations. The force constant for k_r(Si-F) we deduce from this relation is 4.65 x 10^5 dynes/cm. It is larger than that of k_r(Si-H) by about the same factor as the ratio of the bond strengths, 2.0 as compared to 1.8. Calculations imploying this force constant, and three-body forces for Si-Si-F deduced from molecules yield the stretching frequency at ∿833 cm^{-1} with only very small differences between the calculations based on 14 atom and 18 atom clusters (see Fig.4). The ratio of F to Si displacements in this mode is 1.25 in accord with that expected from the mass ratio (1.22). The same calculations yield three in-band resonance modes with appreciable ir-strength, shear-type modes at 550 cm^{-1} and 360 cm^{-1} and a bending mode at 240 cm^{-1}. The frequencies and relative strengths are in good agreement with recent experimental results [33], where absorptions are reported at 212 cm^{-1}, 350 cm^{-1} and 515 cm^{-1}. We are currently performing additional calculations to further clarify the nature of these resonance modes.

We now consider the vibrations that have been attributed to configurations with more than one F. The Hitachi [16,18] and MPI [17] groups both agree that the symmetric and asymmetric stretching vibrations of an SiF$_2$ group are at 828 and 925 cm^{-1}, respectively. We take these as isolated SiF$_2$ groups, defined in the context of Fig.9(a). The Hitachi group further assigns the

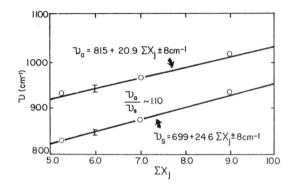

Fig.8 Induction relationships for SiF$_2$

$\nu_a = 815 + 20.9 \, \Sigma X_j \pm 8 cm^{-1}$

$\dfrac{\nu_a}{\nu_s} \sim 1.10$

$\nu_s = 699 + 24.6 \, \Sigma X_j \pm 8 cm^{-1}$

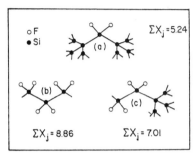

∘ F
• Si

$\Sigma X_j = 5.24$

(a)

(b)

(c)

$\Sigma X_j = 8.86$

$\Sigma X_j = 7.01$

Fig.9 Local arrangements for SiF$_2$ groups including electronegativity sums: (a) isolated group, (b) a mid-polymer group, and (c) a terminal polymer group

vibrations at 870 and 965 cm^{-1} to "another SiF$_2$ configuration". We take this to be a mode of a mid-polymer SiF$_2$ group in an (SiF$_2$)$_n$ chain segment. These assignments are supported by the induction relations shown in Fig.8. We have derived relations similar to (1) for the stretching frequencies of the symmetric and asymmetric groups of SiF$_2$ using molecular data and the assignments proposed above. These turn out to be self-consistent, and give additional insight into the nature of the 1015 cm^{-1} mode. According to the result shown in the figure this is an aysmmetric mode of the (SiF$_2$)$_n$ polymer, associated with an SiF$_2$ group whose neighbors are also SiF$_2$ groups, Fig.9(b). In this context the 1015 cm^{-1} mode in a-Si:F plays the same role as the 845 cm^{-1} in a-Si:H. It is the feature that indicates a high degree of polymerization.

6. Additional Comments

We have also performed preliminary calculations for SiCl and SiCl$_2$ groups. For these two groups the mass of the terminal atom Cl, is greater than that of the Si-atom. This manifests itself in (1) increased Si-motion in all Si-Cl vibrations, (2) more modes close to or in resonance, and finally a larger splitting between the asymmetric and symmetric stretching modes of SiCl$_2$ as compared to SiF$_2$ (see Table 6). The lower bond-strength, bond-polarity and smaller Π-bonding contribution predict a smaller force constant for k_r(Si-Cl) as compared to k_r(Si-F) as well as smaller induction effects. Table 8 gives our predictions for the frequencies of the various bond-stretching modes. These all lie between 500 and 600 cm^{-1}. The SiCl mode is very close to the host modes of the a-Si network. From the calculated frequency of 508 cm^{-1} we can not accurately predict its character, either local or resonance. On the other hand the asymmetric modes of isolated

Table 8 SiCl stretching frequencies

Group	Calculated Frequency (cm^{-1})
Si Cl	508
Si Cl_2	500,580
$[Si\ Cl_2]_n$	545,598

$SiCl_2$ and polymeric $(SiCl_2)_n$ are most likely local modes. More work is clearly in order, experiments to determine where the various stretching modes occur, as well as more reliable calculations.

In summary, we have used chemical bonding arguments based on induction [19] to yield assignments for the bond-stretching vibrations of the various Si-H and Si-F groups. For the SiH and SiF local groups, these calculations support the assignments based on mode counting as well as other considerations relating to the total hydrogen and fluorine content. The calculations also indicate that all of the remaining features in the spectra can be attributed to either SiX_2 or $(SiX_2)_n$, with the polymeric components dominating at higher concentrations of H and F. In this context there is no necessity to invoke appreciable contributions from SiX_3 groups as have been suggested in other studies [4,7].

We are currently performing calculations which combine the isolated cluster and Cluster Bethe Lattice methods. This approach allows the use of realistic force fields, such as those discussed above, but also provides for reasonable boundary conditions on all atoms of a central cluster. Of particular interest in our current work are the line-widths of the various vibrations, and the strengths, frequencies and widths of the in-band resonances.

Acknowledgement

The author acknowledges support for a portion of these works through SERI Subcontract No. HZ-0-9238 under EG-77-C-01-4042. He further acknowledges the contributions of Margaret Memory in the computer calculations relating to the vibrational modes of the clusters.

References

1. A.J. Lewis, G.A.N. Connell, W. Paul, J.R. Pawlik, R.J. Temkin: In AIP Conference Proceedings 20 (American Inst. of Phys., New York, 1974) p. 27
2. A Madan, S.R. Ovshinsky: J. Non-Cryst, Solids 35-36, 171 (1980)
3. W.E. Spear, P.L. LeComber: Solid State Commun. 17, 1193 (1975)
4. M.H. Brodsky, M. Cardona, J.J. Cuomo: Phys. Rev. B16, 3556 (1977)
5. J.C. Knights, G. Lucovsky, R.J. Nemanich: Philos. Mag. B37, 467 (1978)
6. C.C. Tsai, H. Fritzsche, M.H. Tanielian, P.T. Graczi, P.C. Persans, M.A. Vesaghi: In Amorphous and Liquid Semiconductors, ed by W.E. Spear (G.G. Stevenson, Dundee 1977) p. 3
7. G. Lucovsky, R.J. Nemanich, J.C. Knights: Phys. Rev. B19, 2064 (1979)
8. E.C. Freeman, W. Paul: Phys. Rev. B18, 4288 (1978)

9. J.C. Knights, F.A. Street, G. Lucovsky: J. Non-Cryst. Solids 35-36, 279 (1980)
10. H. Wieder, M. Cardona, C.R. Guarnieri: Phys. Stat. Sol. B92, 99(1979); Y. Katayama, T. Shimada, K. Usami, S. Ishioka: Jap. J. Appl. Phys. 19, 115 (1979)
11. C.C. Tsai: Phys. Rev. B19, 2041 (1979)
12. D.K. Paul, J. Blake, B. vonRoedern, R.J. Collins, G. Modell, W. Paul: In Proc. 15th Int. Conf. on Physics of Semiconductors (in press)
13. S. Oguz, R.W. Collins, M.A. Paesler, W. Paul: J. Non-Cryst. Solids 35-36, 231 (1980)
14. A. Madan, S. R. Ovshinsky, E. Benn: Philos. Mag. B40, 259 (1979)
15. H. Matsumura, Y Nakagome, S. Furukawa: Appl. Phys. Lett. 36 (439) 1980
16. T. Shimada, Y. Katayama, S. Horigome: Jap. J. Appl. Phys. 19, 265 (1980)
17. C.J. Fang, L. Ley, H.R, Shanks, K.J. Gruntz, M. Cardona: Phys. Rev. B15 (in press)
18. T. Shimada, Y. Katayama: In Proc. 15th Int. Conf. on Physics of Semi-conductors (in press)
19. G. Lucovsky: Solid State Commun. 29, 571 (1979)
20. F.A. Cotton, G. Wilkinson: Advanced Inorganic Chemistry (Wiley, New York 1972)
21. J.E. Huheey: Inorganic Chemistry 2nd Ed. (Harper and Row, New York 1978)
22. R.T. Sanderson: Chemical Periodicity (Reinhold, New York 1960)
23. R. Alben, D. Weaire, J.E. Smith, Jr., M.H. Brodsky: Phys. Rev. B11, 2271 (1975)
24. S.C. Shen, C.J. Fang, M. Cardona, L. Genzel: Phys. Rev. B15 (in press)
25. J.D. Joannopoulos, F. Ynduruin: Phys. Rev. B10, 5164 (1974)
26. W.B. Pollard, J.D. Joannopoulos: Phys. Rev. B15 (in press)
27. A.L. Smith, N. C. Angelotti: Spectrochimica Acta 15, 412 (1959)
28. H.W. Thompson: Spectrochimica Acta 16, 238 (1960)
29 R.J. Bell, P. Dean: Discuss. Faraday Soc. 50, 55 (1970)
30. P.N. Keating: Phys. Rev. 145, 637 (1966)
31. W. Paul: Solid State Commun. (in press)
32. G. Lucovsky: Solar Cells (in press)
33. C.J. Fang, K.J. Gruntz, L. Ley, M. Cardona, F.C. Demmond, G. Muller, S. Muller, S. Kalbitzer: J. Non-Cryst. Solid 35-36, 255 (1980)

Photo-Induced Phenomena in Amorphous Semiconductors

Kazunobu Tanaka

Electrotechnical Laboratory
1-1-4, Umezono, Sakura-mura
Niihari-gun, Ibaraki 305, Japan

1. Introduction

Amorphous solids, chalcogenide glasses as well as tetrahedrally-coordinated
amorphous materials, are characterized as a phase which is structurally rigid
but possesses no long-range order, due to which their optical, electrical and
structural properties are generally smeared out in contrast to those of
crystalline counterparts. It seems to be one of the reasons why some people
hesitate to get into this field of amorphous semiconductors.

On the other hand, from the viewpoints of thermodynamics, an amorphous
solid is in a non-equilibrium state, therefore, its structure and bond config-
rations are not fixed but can be changed, sometimes reversibly, not only by
thermal treatment but also by light irradiation. This fundamental difference
based on the thermodynamics between amorphous and crystalline solids makes
the subject of amorphous semiconductors more interesting.

In this lecture I should like to give a brief description of photo-induced
effects observed in several amorphous chalcogenides such as As_2S_3, As_2Se_3 and
GeS_2. The effects are phenomenologically separated into two different phenomena; a reversible and an irreversible changes. The emphasis will be placed
on the reversible photostructural change accompanying optical change (so-
called photodarkening) induced by successive cycles of band-gap illumination
and thermal annealing near the glass transition temperature. The results of
X-ray diffraction, Raman scattering and volume change measurements will be
given as the direct evidences for the reversible photostructural change,
being discussed in relation with the reversible shift of optical absorption
edge (photodarkening).

It will be shown through whole discussion that the phenomenon is "unique
to amorphous phase". The mechanism underlying the phenomenon has not yet
been well understood, but I shall try to describe possible mechanisms presented by several groups independently so far.

2. Material and Structure

Photo-induced optical and structural changes have been observed in various
kinds of chalcogenide glasses; a-Se, a-As_2S_3, a-As_2Se_3, a-$GeSe_2$, a-GeS_2, a-
$As_4Se_5Ge_1$, a-As-S-Se-Ge and so on. The key elements of these glasses are
the chalcogen, S and/or Se [1]. In this note I shall take a-As_2S_3 as a typical chalcogenide glass showing photo-induced structural change, and briefly
describe its atomic and electronic structures before proceeding to the details of the phenomena.

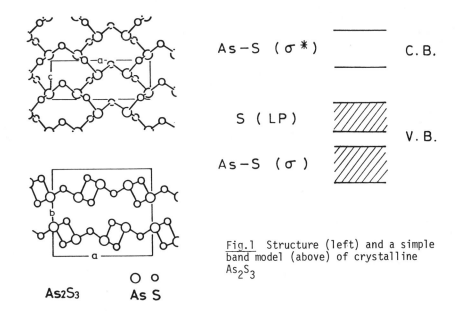

As—S (σ*)   C. B.

S (LP)

V. B.

As—S (σ)

Fig.1 Structure (left) and a simple band model (above) of crystalline As_2S_3

As_2S_3 ○ ○   As S

Crystalline As_2S_3 has a monoclinic structure formed by 3-fold coordinated As and 2-fold coordinated S atoms, as shown in Fig.1 [2]. It involves two essentially different kinds of chemical bonds; a strong covalent bonding (As-S bond) and a weak van-der-Waals-like bonding between layers.

The above nearest-neighbour coordination associated with covalent bonds is essentially retained even in the disordered network of As_2S_3 glass, which has been confirmed by X-ray diffraction [3], Raman scattering [4] and nuclear quadrupole resonance studies [5].

In contrast to the nearest-neighbour ordering we have less information on the second or third nearest-neighbour configurations. What we have known at present is that the absence of long-range order produces a small amount of homobonds such as As-As or S-S and, at the same time, a much lower concentration (10^{17}-10^{18}/cm^3) of bonding defects in terms of D^+ and D^- or C_3^+ and C_1^- in glassy network·[6 - 9].

The most characteristic feature, common to chalcogenide glasses, is the electronic structure originated in p-like lone pair electrons of chalcogen atoms. A sulphur atom has four outer p electrons, two of which form two covalent bonds (As-S bonds) with p electrons from neighbouring As atoms and the remaining two are localized on the chalcogen atom as lone-pair electrons.

A very simple band structure model for As_2S_3 glass is also shown in Fig.1. It should be noted that there is no essential difference in a band structure between crystalline and glassy As_2S_3. The top of the valence band is formed out of the lone-pair (LP) non-bonding p-states and the two bonding (σ) p-states lie in a lower energy [10]. The presence of these high energy LP p-

105

states is associated with the defects models, the valence-alternation-pair (VAP in terms of C_3^+ and C_1^-) presented by KASTNER, ADLER and FRITZSCHE [9] and the charged dangling bonds (in terms of D^+ and D^-) by STREET and MOTT [10].

The above LP electrons as well as a low coordination number of the chalcogen atom are considered as an origin of strong electron-phonon coupling in the chalcogenide glass system, and give a large flexibility to various local atomic configurations.

3. Irreversible Structural Change

Optical and structural changes induced by band-gap illumination are observed not only in evaporated (or sputtered) thin films but also melt-quenched bulk glasses [11-13]. In many cases these changes are reversible in the sense that an annealing treatment for some time at a temperature higher than the illumination temperature can restore the initial structural state and optical properties.

However, in the case of thin films, the initial structure is often much different from that of melt-quenched glasses and undergoes irreversible structural changes, when illuminated or annealed, which can not be regained by any simple combination of annealing and further illumination.

In the early stage DENEUFVILLE and his coworkers have interpreted the irreversible process of as-evaporated As_2S_3 film in terms of a polymerization of the As_4S_6 molecular glass through the detailed X-ray diffraction analyses

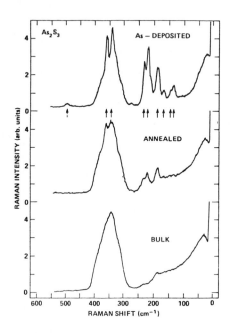

Fig.2 The Raman spectra for the as-deposited, annealed and bulk-glass forms of As_2S_3 (from NEMANICH et al.) [6]

106

[14]. Recent results of Raman scattering and NQR have given more detailed information on the structure of as-deposited As_2S_3 films, and suggested a possible mechanism for the irreversible process.

Figure 2 shows the Raman spectra for the as-deposited, annealed, and bulk-glass forms of As_2S_3, which have been reported by NEMANICH and his coworkers [6]. The several sharp peaks of as-deposited film decrease or disappear irreversibly after annealing just below the glass transition temperature, indicating that the gross structural change takes place. This irreversible change is partially induced by band-gap illumination. The data analyses indicate that all the sharp spectra features of as-deposited As_2S_3 except one at 490 cm^{-1} are from As-As bonds, suggesting the presence of As_4S_4 molecules [6].

Very recently, TREACY and his coworkers have reported the NQR (nuclear quadrupole resonance) measurements [15]. In Fig.3 the NQR absorption spectra at 77 K are shown for bulk As_2S_3 (dotted line) and 300-K substrate (as-evaporated) films of As_2S_3 before (dashed line) and after (solid line) irradiation. The NQR frequencies observed in various crystalline forms of the As-S system are shown as solid lines at the top of the figure. As shown in the figure, the NQR absorption spectrum of the as-deposited As_2S_3 changes drastically by 5145-A irradiation. According to their interpretation 300-K substrate films of As_2S_3 posseses AsS_3 pyramidal units, in addition to As_4S_4 molecular units, which lack the local 2-D order present in the bulk As_2S_3. Therefore, it has been tentatively speculated that irreversible photo-induced structural changes tend to produce order similar to that which occurs in the bulk [15].

The above irreversible structural change always accompanies optical changes, an increase in the refractive index and a red-shift of the optical absorption edge [1][14], which has not yet been well understood.

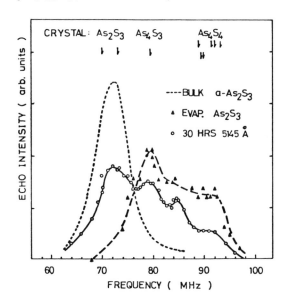

Fig.3 Pulsed ^{75}As NQR (nuclear quadrupole resonance) echo intensity as a function of frequency for the as-evaporated (300-K substrate), 30-hrs illuminated (5145 A), and bulk-glass forms of As_2S_3 (from TREACY et al.)[15]

4. Reversible Photo-Induced Changes

Reversible photodarkening and photostructural change have been observed in well-annealed evaporated or sputtered films as well as melt-quenched bulk glasses [12]. In this section I should like to give our detailed experimental results obtained mainly for a well-annealed evaporated As_2S_3 (EV-As_2S_3) and a melt-quenched As_2S_3 (MQ-As_2S_3).

4.1 Optical Changes (Photodarkening)

Many groups have reported reversible photodarkening in thin films of chalcogenides [16-19]. The optical change termed "photodarkening" has been characterized by a nearly parallel shift of the absorption edge [17-19]. This optical change is induced even in a melt-quenched glass [11].

Figure 4 shows the reversible photodarkening observed both in EV- and MQ-As_2S_3 after the repeated cycles of illumination/annealing at 170°C [11]. As shown in the figure, both samples show the similar quantity of the edge shift, while no change in crystalline As_2S_3. It indicates that photodarkening is an "unique" phenomenon originated from disordered structure.

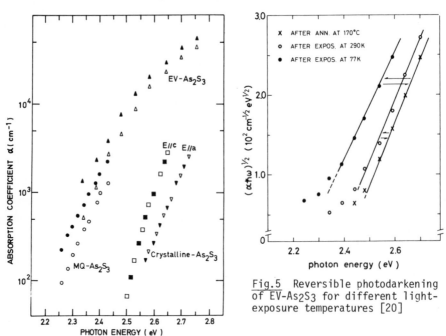

Fig.5 Reversible photodarkening of EV-As_2S_3 for different light-exposure temperatures [20]

Fig.4 The absorption edge spectra of EV-As_2S_3, MQ-As_2S_3 and crystalline As_2S_3 before (open symbols) and after (solid symbols) the band-gap illumination [11]

Generally, $|\Delta E_o|$ (edge shift) increases with a decrease in the temperature at which the sample is exposed to light. Figure 5 shows the photodarkening in EV-As_2S_3 for different exposure temperatures [20]. It should be noted all the spectra were measured at room temperature, which implies that the photodarkening includes some kind of a nonlinear process.

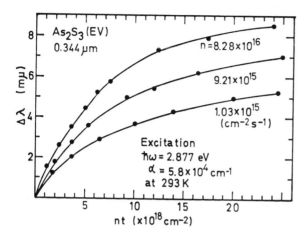

Fig.6 Edge shift as a function of nt product for different incident photon flux n ($\Delta\lambda$ is a shift of wavelength corresponding to $\alpha = 3 \times 10^4$ cm^{-1})

Such a nonlinearity is reflected more clearly in the break of the reciprocity law between incident photon flux n and exposure time t. It has long been believed that ΔE_o is affected only by $\overline{nt}$ product [16], but recently, we have observed that $|\Delta E_o|$ gets larger gradually with an increase in n under the condition of a constant $\overline{nt}$ product. The results for EV-As$_2$S$_3$ is shown in Fig.6 [13], which indicates that the reciprocity law of n vs. t does not precisely hold in the photodarkening process. Detailed analyses of the data have shown that photodarkening process cannot be interpreted within the framework of the first-order reaction even if the backward thermal restoration (proportional to $\nu\exp[-V/kT]$) is taken into account [21]. This nonlinearity has been tentatively ascribed to a strong interaction between neighbouring structural change units (neighbouring two double-well potentials to be mentioned below).

4.2 Structural Changes

First direct evidence for the existence of a reversible photostructural change has been given using X-ray diffraction measurements [20]. Figure 7 shows the reversible changes in diffracted X-ray intensity curve of EV-As$_2$S$_3$ which occur simultaneously with optical changes shown in Fig.5. The first diffuse peak is slightly shifted to higher angles as a result of exposure, and its intensity is lowered, while the minimum at $2\theta \simeq 20°$ increases. Such a tendency is enhanced for the case of the exposure to light at 77 K, being consistent with the exposure temperature dependence of optical change in Fig.5. We have also observed a similar reversible change in EV-As$_4$Se$_5$Ge$_1$ [19].

The common feature is the decrease in amplitude difference between the first maximum and minimum of the curve, approximately corresponding to a decrease in the amplitude of the oscillatory function,

$$NI\Sigma\Sigma f_i f_j [\sin(sr_{ij})/sr_{ij}], \tag{1}$$

involved in the Debye formula. Namely, photostructural change can be characterized with an increase in randomness of atomic configuration.

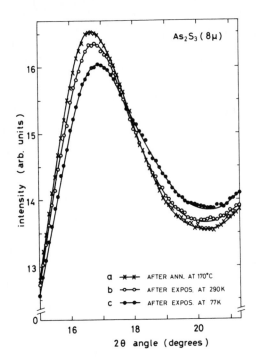

As₂S₃(8μ)

a ✕✕ AFTER ANN. AT 170°C
b ○○ AFTER EXPOS. AT 290K
c ●● AFTER EXPOS. AT 77K

Fig.7 Photo-induced changes in X-ray diffraction curve of EV-As$_2$S$_3$ for different light-exposure temperatures [20]

UTSUGI and MIZUSHIMA have observed a reversible change in the far-infra-red absorption spectra of amorphous Se-Ge film, and concluded that photo-irradiation induces a large fluctuation in the Se-Ge bond angle [22]. KUMEDA and his coworkers have also suggested an increase in randomness through ESR study on amorphous As$_2$Se$_3$:Mn system [23].

Another evidence for the reversible photostructural change has been obtain-ed by the thin film dilatometry using a mechanical method, detailed procedures of which are given in Ref.[24]. Photo-induced thickness changes of EV- and MQ-As$_2$S$_3$ for different exposure temperatures are summarized in Table 1.

Table 1 The reversible photo-induced change of the thickness of EV- and MQ-As$_2$S$_3$ [13]

	Exp. Temp. [K]	Thickness d[μm]	Thickness change Δd[A]	Fraction Δd/d
EV-As$_2$S$_3$	293	0.96	40	0.42
	293	2.10	80	0.38
	293	4.20	160	0.38
MQ-As$_2$S$_3$	293	16.7	850	0.51
EV-As$_2$S$_3$	163	4.20	298	0.71

As seen in the table, a fractional change $\Delta d/d$ has almost no dependence on d, indicating that the change should be ascribed to a volume effect. It is noted that $\Delta d/d$ of MQ-As_2S_3 nearly coincides with that of EV-As_2S_3 within the estimation error. The value of $\Delta d/d$ is considered to be that of fractional volume change $\Delta V/V$, to a first approximation, since no discernible change of the area of each sample surface was observed.

Lower exposure temperature produces larger changes, which is a common tendency also observed in the results shown in Figs.5 and 7.

4.3 Spectral Response

A crucial characterization of any photo-induced effect involves the determination of its spectral response. In the Edinburgh conference, we reported the first data on the spectral response of the reversible photodarkening [12].

Figure 8 shows the edge shift E_0 as a function of photon energy of exciting light observed in As_2S_3 at 293 K which is accurately normalized to the number of absorbed photons. The spectrum extending over a wide range of photon energies was obtained from the measurements on EV- and MQ-As_2S_3 samples under the conditions determined by the following relations;

$$\alpha d = \text{const.} \tag{2}$$

$$n\alpha(1 - R) = \text{const.}, \tag{3}$$

where $\alpha = \alpha(h\nu)$ is the absorption coefficient, d the sample thickness, n the incident photon flux and R the reflectivity at normal incidence. ΔE_0 tends to a constant value for $h\nu > 2.4$ eV ($\alpha > 10^3$ cm^{-1}), while it falls off steeply in low photon energies ($h\nu < 2.4$ eV).

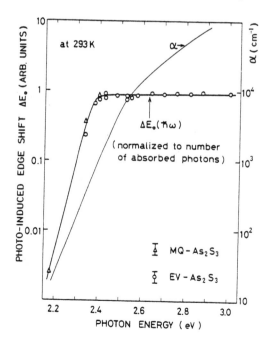

Fig.8 Spectral response of photodarkening in a-As_2S_3 [12]

It is apparent that the photodarkening is not easy to be detected in MQ-bulk sample whose thickness $\underline{d} > 100$ μm, since the phenomenon is efficiently induced by light whose penetration depth is less than 10 μm.

From the data in Fig.8 it is clear that the photodarkening is essentially induced by the interband optical absorption. It is interesting to compare the data with the photoluminescence (PL) excitation spectrum. The slow-decay PL excitation spectrum normalized to absorbed photon flux shows a constant relative quantum efficiency for low hν for which $\alpha < 10^2$ cm^{-1} and decay gradually in high hν region, which is nearly complementary to ΔE_o(hν) of photodarkening [25]. On the other hand, PC spectral response is similar to ΔE_o(hν) except that the shoulder of the spectrum is shifted to higher hν than that of ΔE_o(hν) [26]. Therefore, it is tentatively speculated that an exciton does not contribute to PC but to the photodarkening.

Table 2 shows the spectral response of photostructural change in terms of $\underline{\Delta d}/\underline{d}$ ($\Delta V/V$) observed for MQ-GeS$_2$ [27]. The value of $\underline{\Delta d}/\underline{d}$ also falls off in low hν region ($\alpha < 10^3$ cm^{-1}).

Table 2 Spectral response of photostructural change in MQ-GeS$_2$ [13]

Excitation	Absorption coefficient	Photon flux	Exposure time	Thickness change
hν[nm]	α[cm^{-1}]	n[cm^2/sec]	t[sec]	$\Delta d/d$
374.5	3.1×10^3	1.52×10^{17}	2.52×10^4	4.7×10^{-3}
392.5	1.1×10^3	1.74×10^{17}	2.88×10^4	1.9×10^{-3}
403	6.0×10^2	2.57×10^{17}	2.88×10^4	5.1×10^{-4}
411	3.8×10^2	9.65×10^{16}	3.60×10^4	$< 6 \times 10^{-5}$

4.4 Photo-Induced Effects at Low Temperature

Photo-induced optical change at low temperature involves two essentially different phenomena; a parallel shift of the absorption edge (i.e., normal photodarkening) and a broad absorption band in the mid-gap region. The latter is caused by the formation of paramagnetic centers (like D^o), therefore, always accompanied by ESR signals ($n_s = 10^{17} - 10^{18}$ cm^{-3}) [7].

Figure 9 shows the photo-induced optical change in EV-As$_2$S$_3$ at 14 K for 4880-Å excitation, which cannot be characterized by a simple parallel shift of the absorption edge. The induced absorption in lower photon energies can be released not only by warming to room temperature but also by irradiating the darkened film with light with photon energy lower than band gap E$_o$, while a portion of the large shift of the absorption edge remains stable even after both procedures [13].

Optical bleaching by 6471-Å light shown in the figure is accompanied by simultaneous disappearance of ESR signal. It is clear that the mid-gap absorption bleached by a 6471-Å light (indicated by symbol "x" in the figure) is qualitatively the same as that observed by BISHOP and his coworkers in MQ-As$_2$S$_3$ although the strength of the induced absorption in the figure (EV-As$_2$S$_3$) is higher by one order of magnitude than that of MO-As$_2$S$_3$.

112

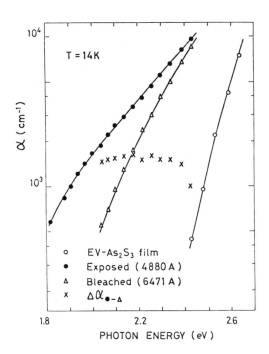

The inducing band lies approximately in the range of $1 < \alpha < 10^4$ cm^{-1}, which is almost the same as PL excitation spectrum, namely, complementary to that of photostructural change.

Recently, SHIMIZU et al. have observed photo-induced optical change accompanying ESR signal in Ge$_{30}$S$_{70}$ film at room temperature. Glass transition temperatures of the Ge-S system are higher by 300 - 400 K than that of As$_2$S$_3$. It is quite possible that mid-gap absorption due to paramagnetic centers in Ge$_{30}$S$_{70}$ remains stable even at room temperature, being superposed on normal photodarkening [34].

Very recently also, BIEGELSEN and STREET have reported that a prolonged exposure to strongly absorbed light induces new metastable defects with a density of 10^{18} - 10^{20} cm^{-3} [28]. These newly created defects accompany ESR signals whose feature is different from that observed by BISHOP et al. But these centers are washed out by warming up to room temperature.

4.5 Mechanisms

In order to suggest possible mechanisms of the phenomenon, I should like to discuss the above results in a unified manner. The discussion will be divided into several broad topics covering the nature and origin of photostructural change and the relationship between optical and structural changes.

4.5.1 Structural Origin (Bulk or Defect?)

Mid-gap absorption as well as slow-decay PL is associated with the defects in terms of "charged dangling bonds" or "VAP's". It might be reasonable to consider that the density of such defects has a strong dependence on a local fluctuation of structure originated from preparation methods. This is the reason why the mid-gap absorption and PL reflect the difference between EV- and MQ- samples [12][25].

On the other hand, no essential difference has been observed between photo-structural change accompanying photodarkening of MQ- vs. EV-samples, as des-cribed above.

Mid-gap absorption involves the optically induced centers of around 10^{17} cm^{-3} which has been estimated from ESR spin density (n_s). Regarding the photo-structural change, the density of local atomic displacements can be tentatively estimated from the experimental data on $\Delta V/V$. By assuming that one local atomic displacement creates a free volume equivalent to the mean atomic volume for a maximum case, we obtain 10^{20} - 10^{21} cm^{-3} as a density.

Comparisons between various properties of photostructural change and mid-gap absorption are summarized in Table 3. Thus it is clear that the origin of the photostructural change is not directly related with the native defects but with bulk-oriented feature in a disordered network, in other words, a quantitative disorder such as fluctuations in bond angle or closed shell dis-tance.

Table 3 Comparison between photostructural change (PSC) and mid-gap absorp-tion and their structural origins [13]

	PSC	Mid-gap absorption
Room temp.	Yes	No
Optical change	$\Delta E_o < 0$	$\Delta \alpha > 0$
ESR	No	Yes
Excit. spectrum	High α	Low α
Density	10^{20}-$10^{21}/cm^3$	10^{17}-$10^{18}/cm^3$
Preparation depend.	Minor	Major
Origin	Bulk	Defects(D^+,D^-)

Figure 10 shows a schematic model of bistable local bonding geometries. A and A' represent local bistable configurations which give two minima in the potential energy of the system as a function of some appropriate local atomic coordinates. A concept of "double well potential" has been originally pre-sented for the linear temperature dependence of a specific heat, in which a tunneling process is assumed [29]. Such a double well condition is originated from structural randomness, since the crystalline counterpart has essentially a single minimum and actually shows neither photostructural change nor photo-darkening.

MURAYAMA et al. have pointed out the importance of the interaction between an electron and lattice vibrations through their experimental results on the fast-decay PL [30]. It is reasonable to consider that a local structural change without bonding defects, like a transition from A to A', is induced by a strong electron-phonon coupling involved in a recombination process of a photo-excited valence electron.

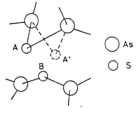

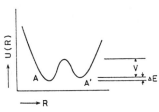

Fig.10 A shematic model of bistable local bonding geometries and corresponding double well potential.

STREET has proposed a self-trapped exciton model in terms of $[D^+, D^-]$ pair for the photostructural change in connection with the recent data on new defect centers mentioned above [28][31]. This is a possible alternative model, but a more detailed study will be required.

4.5.2 Compositional Trends of the Phenomenon

Photostructural change and photodarkening observed in various stoichiometric chalcogenide glasses are summarized in Table 4 [13]. The A - A' transition of atomic configuration shown in Fig.10 involves implicitly not only a change in the distance between closed shells (A - B) but also a change in bond angle. Actually, photoinduced increase in fluctuation in Se-Ge bond angle has been directly observed in amorphous sputtered Se-Ge film by UTSUGI and MIZUSHIMA [22], although TREACY et al. have suggested from the NQR studies that the photodarkening in a-As_2Se_3 accompnies no gross change in a bond angle [15].

Such a bending flexibility is strongly correlated with an ionicity of chemical bond which may possibly reflect the quantitative difference in $\Delta E_o/E_o$ between four stoichiometric chalcogenide binary alloys in the table.

For simplicity, two quantities are picked up as important factors affecting photostructural change: One is a difference in electronegativities between constituent elements (X_A-X_B) giving a measure of A-B bond ionicity, and the other an average coordination number $\bar{n}$ from the viewpoints of structural flexibility.

Each calculated value of (X_A-X_B) and $\bar{n}$ is listed for four binary compounds in the table. As seen in the table, $\Delta E_o/E_o$ and $\Delta V/V$ indicate larger values for higher ($\overset{*}{X}_A-X_B$) and/or for lower $\bar{n}$, which is consistent with the above discussion.

As has been verified by various structural studies, the stoichiometric glass materials listed in the table are chemically ordered, that is, involve a less amount of homobonds in their networks. In particular, MQ-GeS_2 glass is well chemically ordered and the concentration of homo-bonds such as Ge-Ge and S-S has been estimated to be less than 0.01 % [27], which means that wrong bonds should not be responsible for a structural origin of the photostructural change.

115

Table 4   Compositional trends of photostructural change and photodarkening[13]

		As_2Se_3	As_2S_3	$GeSe_2$	GeS_2
E_0[eV]		1.76	2.41	2.20	3.15
ΔE_0		-0.015	-0.045	-0.032	-0.075
$|\Delta E_0|/E_0$		8.52×10^{-3}	1.87×10^{-2}	1.46×10^{-2}	2.38×10^{-2}
$X_A - X_B$		0.22	0.30	0.44	0.52
$\bar{n}$		2.40	2.40	2.67	2.67
	MQ	3.3×10^{-3}	6.1×10^{-3}	3.7×10^{-3}	4.7×10^{-3}
$\Delta V/V$	EV	4.4×10^{-3}	6.0×10^{-3}		
	SP				
Change in X-ray Diff.	MQ		Yes		
	EV	Yes	Yes		
Excitation Spect.		$\alpha \geq 10^3 cm^{-1}$	$\alpha \geq 10^3 cm^{-1}$		$\alpha \geq 10^3 cm^{-1}$

$X_A - X_B$: Difference in electronegativity between constituents of AB alloys
$\bar{n}$: Average coordination number

4.5.3 Optical Gap and Structural Randomness

From a phenomenological point of view, the optical and structural experimental data for various chalcogenides described above clearly show that the reversible photo-induced edge shift ΔE_0 is attributed to structural change characterized with an increase in randomness.

From the data on ΔE_0 and $\Delta V/V$ listed in Table 4, the fractional volume coefficient of E_0 can be estimated. It is interesting to compare the obtained result with that determined from the pressure studies [32].

Table 5 shows the comparison between pressure- and photo-induced effects. Striking differences between two effects are observed both in a sign and a magnitude of the coefficient, suggesting that ΔE_0 cannot be discussed only by $\Delta V/V$. A change in a structural randomness should be taken into account as an important factor affecting photo-induced effect.

The average of the broadening of valence and conduction bands increases as structural randomness increases [32], to a first approximation, resulting in a decrease in E_0. This is qualitatively consistent with the present experi-

Table 5 Volume coefficient of the optical absorption edge (comparison between pressure-induced and photo-induced effects) [13]

| | Pressure-induced | | Photo-induced |
	$(\partial E_o/\partial P)_T$ $[10^{-5} eV\ bar^{-1}]$	$V(\partial E_o/\partial V)_T$ [eV]	$V(\partial E_o/\partial V)_T$ [eV]
MQ-As_2S_3	-1±0.2	1.1±0.2	-7.4±0.8
MQ-As_2Se_3	-1.3±0.1	2.0±0.2	-4.5±0.5

mental observation. Therefore, as a more general form, ΔE_o can tentatively
be described as

$$\Delta E_o = [V(\partial E_o/\partial V)](\Delta V/V) + [D(\partial E_o/\partial D)](\Delta D/D), \tag{4}$$

where D is a quantity representing a degree of structural randomness. As
$V(\partial E_o/\partial V) > 0$ and $D(\partial E_o/\partial D) < 0$, the first and the second terms are competing
factors in the case of photo-induced effect. Pressure-induced edge shift is
explained mainly by the first term (volume effect), while the second term
(randomness) overcomes the effect of the first term in the case of photo-
induced phenomenon.

In a recent work, ABE has given a theoretical study on the interband optical transitions in a random system taking into account the effect of interband correlation of randomness. His tentative results have suggested that
a negative correlation (antiparallel potential fluctuations) causes a red-
shift of the absorption edge [33]. For further discussion, more systematized
characterization of structural change as well as interpretation of absorption
itself will be required.

5. Summary

Reversible and irreversible photo-induced phenomena observed in chalcogenide
glasses are reviewed with a main focus on the former. The following items
are emphasized: (1) The photodarkening is caused by the simultaneous photostructural change characterized with an increase in structural randomness.
(2) The phenomenon is unique to amorphous phase, and commonly observed both
in evaporated films and in melt-quenched glasses. (3) The origin of the
photostructural change is not a defect such as a native dangling bond or a
VAP but a quantitative disorder inherent to amorphous bulk network. (4)
Compositional trends of the phenomenon are discussed with a bond ionicity & an
average coordination number in connection with double well potential model.
(5) A relationship between E_o and structural randomness is discussed in
comparison with the pressure effect. (6) Detailed mechanism underlying the
phenomenon is still an open question.

References

1. J.P. deNeufville: Optical Properties of Solids - New Developments, ed. by B.O. Seraphin (North-Holland Pub.Co., Amsterdam 1975) p.437
2. R. Zallen: Optical and Electrical Properties (Physics ans Chemistry of Materials with Layered Structures, vol.4) ed. by P.A. Lee (D.Reidel Pub. co., Dordrecht 1976) p.231
3. A.J. Leadbetter, A.J. Apling: J. Non-Cryst.Solids 15, 250(1974)
4. G. Lucovsky, R.M. Martin: J. Non-Crystalline Solids 8-10, 185(1972)
5. M. Rubinstein, P.C.Taylor: Phys.Rev.Letters 29, 119(1972)
6. R.J. Nemanich, G.A.N. Connell, T.M. Hayes, R.A. Street: Phys.Rev. B18, 6900(1978)
7. S.G. Bishop, U. Strom, P.C. Taylor: Phys. Rev. Letters 34, 1346(1975)
8. R.A. Street, N.F. Mott: Phys. Rev. Letters 35, 1293(1975)
9. M. Kastner, D. Adler, H. Fritzsche: Phys. Rev. Letters 37, 1504(1976)
10. M. Kastner: Phys.Rev. B7, 5237(1973)
11. H. Hamanaka, K. Tanaka, S. Iizima: Solid State COMMUN. 23, 63(1977)
12. K. Tanaka: Proc. 7th Intern. Conf. Amorphous and Liquid Semiconductors, Edinburgh, ed. by W.E. Spear (1977)p.787
13. K. Tanaka: J. Non-Crystalline Solids 35&36, 1023(1980)
14. J.P. deNeufville, S.C. Moss, S.R. Ovshinsky: J. Non-Crystalline Solids 13, 191(1973/74)
15. D.J. Treacy, U. Strom, P.B. Klein, P.C. Taylor, T.P. Mratin: J. Non-Crystalline Solids 35&36, 1035(1980)
16. J.S. Berkes, S.W. Ing.,Jr., W.J. Hillegas: J. Appl. Phys. 42, 4908(1971)
17. S.A. Keneman: Appl. Phys. Letters 19, 205(1971)
18. T. Igo, Y. Toyoshima: Proc. 5th Conf. (1973 International) Solid State Devices, Tokyo (1973)p.106.
19. K. Tanaka, S. Iizima, K. Aoki, S. Minomura: Proc. 6th Intern. Conf. Amorphous & Liquid Semiconductors, Leningrad, ed. by B.T. Kolomiets (Nauka, 1976)p.442
20. K. Tanaka: Appl. Phys. Letters 26, 243(1975)
21. H. Hamanaka, K. Tanaka:Unpublished
22. Y. Utsugi, Y. Mizushima: J. Appl. Phys. 49, 3470(1978)
23. M. Kumeda, Y. Nakagaki, M. Suzuki, T. Shimizu: Solid State Commun. 21, 717(1977)
24. H. Hamanaka, K. Tanaka, S. Iizima: Solid State Commun. 19, 499(1976)
25. R.A. Street: Adv. Phys.25, 397(1976)
26. H.K. Rockstad: J. Non-Crystalline Solids 2, 192(1970)
27. H. Hamanaka, K. Tanaka : Solid State Commun. 33, 355(1980)
28. D.K. Biegelsen., R.A. Street : Phys. Rev. Letters 44, 803(1980)
29. P.W. Anderson, B.I. Haiperin, C.M. Varma : Philos. Mag. 25(1971)
30. K. Murayama, h. Suzuki, T. Ninomiya : J. Non-Crystalline Solids 35&36, 915(1980)
31. R.A. Street : Proc. 7th Intern. Conf. Amorphous & Liquid Semiconductors, Edinburgh, ed. by W.E. Spear (1977)p.509
32. M. Kastner : Phys. Rev. B7, 5237(1973)
33. S. Abe : Proc. Intern. Conf. Physics of Semiconductors, Kyoto, (1980) in press
34. T. Shimizu, M. Kumeda, I. Watanabe, Y. Nakagaki: Solid State Commun. 27, 223(1978)

Theory of Electronic Properties of Amorphous Semiconductors

Fumiko Yonezawa* and Morrel H. Cohen

*Research Institute for Fundamental Physics, Kyoto University
Kyoto 606, Japan

James Franck Institute, University of Chicago
Chicago, IL 60637, USA

Abstract

A typical way is shown to sort out important kinds of disorder found in
a tetrahedrally-bonded elemental amorphous semiconductor such as Si or Ge.
Generally, they are classified in quantitative and topological disorder.
The quantitative disorder includes bond-length, bond-angle, and dihedral-
angle variation, while the topological disorder includes the ring statistics. Using a tight-binding Hamiltonian for Si, we investigate the
relationship between the kinds of disorder and the electronic structures
of amorphous Si such as the energy bounds and the behaviour of the band
tails. Extensions of the results thus obtained enable one to understand
semiquantitatively the differences in the density of states between
crystalline and amorphous Si as well as predict variations among different
amorphous semiconductors.

1.   Introduction

A goal of solid state physics is to predict, from the macroscopically measured physical quantities of a solid, the rules governing the microscopic
world on atomic and electronic scales. This is true both for ordered crystals and for disordered systems such as amorphous solids and liquids. For
the last 40 or 50 years solid state physics has made remarkable progress in
the study of ordered crystals. On the other hand, the theoretical understanding of disordered systems is still largely undeveloped. The important
point in studying amorphous semiconductors, therefore, is to make clear how
far they are analogous to, as well as different from, crystals and other
kinds of disordered systems. In other words; what aspects do amorphous
semiconductors have in common with crystals and other disordered systems,
what is typical and characteristic of amorphous semiconductors, where do
these characteristic features of amorphous semiconductors come from, why are
they interesting and important and how can they be used?

   Our purpose in this paper is to investigate amorphous semiconductors by
regarding them as one class of disordered systems. Our task is again to
connect the microscopic world with the macroscopic world. In particular,
we would like to demonstrate how these problems could be dealt with by reporting a series of theoretical investigations on the electronic properties
of tetrahedrally-bonded semiconductors such as amorphous silicon (a-Si) and
amorphous germanium (a-Ge) [1-5].[1]

---

[1]   A part of the present paper is based on work carried out in collaboration with Dr. Jasprit Singh and Prof. Hellmut Fritzsche.

With the aforementioned questions in mind, we first see in §2 how various disordered systems may be classified. The implications of short-range order and long-range disorder are considered in §3. For later convenience, a brief review is given in §4 to describe what is known about the spectral bounds and band tails of binary alloys and related materials. In §5, we show what parameters are needed to describe the atomic correlations in a-Si. In §6, we study how far we can go with the simplest reasonable model Hamiltonian for a-Si and other four-fold materials and in §7 we examine how much more information could be obtained if the simplest model were improved to include more interactions. Further discussion will be given in the last section, §8.

2. Classification

The first step towards the goal explained in §1 is to define the amorphous state. In a recently published book titled "Amorphous Semiconductors" Brodsky says, "It is easier to define the amorphous state by stating what it is not than precisely specifying what it is"[6]. This situation is common to various kinds of disordered systems. It is most clearly reflected in the terms describing these systems. They are often expressed in negative terms such as *disordered, aperiodic, non*-crystalline, and *amorphous*. Therefore, the most efficient approach would be to state how a "well-defined" ordered system may be distorted to yield a certain class of disordered systems. In this context, it is useful to quote Ziman from his latest book "Models of Disorder": he says, "Disorder is not mere chaos; it implies defective order" [7]. Thus, what we ought to do is to specify how defective a given dis-ordered system is.

One example of the classification of disordered systems is given in Table 1. The schematic pictures in Figs.1 and 2 facilitate understanding of our classification. Needless to say, the disorder or defective order occurs on the microscopic scale. To be more precise, the distributions of atoms are spatially disordered. Disorder always breaks in some way the perfect peri-odicity of an ordered crystal.

Cellular disorder is found in substitutional alloys and mixed crystals, where a crystal lattice is still present but we can no longer say with precision which type of atom is to be found on each lattice site.

In structural disorder, found in liquid and amorphous metals and in topological disorder, found in amorphous semiconductors, 'no ghost of a crys-tal lattice' remains. The positions of atoms are no longer regular in these systems. We classify topological disorder as a category different from structural disorder since the former is typical of covalently-bonded systems. The connectivity of the covalent bonds (which define a random network) plays an essential role in determining some of the physical properties of these systems, as occurs in amorphous semiconductors. The positions of the atoms are disordered, and long-range order (LRO) is naturally absent. It should be noted however that short-range order (SRO) is maintained to a consider-able extent. The bond lengths and angles fluctuate narrowly around respec-tive average values corresponding to those found in ordered crystals. The coordination number of each atom is the same as that found in a crystal except for occasional deviations associated with specific defects.

Table 1 Classification of Spatial Disorder

[I] **Crystals** (ordered systems, periodic systems, (nonrandom systems))

[II] **Noncrystalline** (disordered systems, aperiodic systems, random systems)
     Materials

```
                              ┌─substitutional disorder
         ┌─Cellular disorder──┼─compositional  disorder
         │                    └─quantitative   disorder
         │
         │                    ┌─positional  disorder ┐  ┌─Equilibrium (eg liquids)
         ├─Structural disorder┴─qualitative disorder │  │
         │                                           ├─>│                       ┌─glass
         │                                           │  └─Nonequilibrium────────┤
         │                                              (amorphous solids) └─non-glassy amorphous solids
         │
         ├─Topological disorder    (random network) ┘
         │
         └─Continuum disorder
```

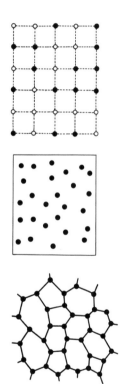

Fig.2 Contour of electron density or potential in (a) a crystal; (b) a system with structural or topological disorder; and (c) a continuum disorder (after Ziman 1979 [7]).

Fig.1 Schematic illustration of various kinds of disorder: (a) cellular disorder, (b) structural disorder and (c) topological disorder.

An important point about amorphous solids is that they are not in the equilibrium state. This is one of the main reasons why amorphous semi-conductors and amorphous metals are so very interesting and at the same time so very difficult to describe. We discuss this problem later in some detail.

It is also appropriate to mention here that the amorphous solids are sometimes divided into glasses and non-glasses; glasses are those amorphous solids which have been cooled from the melt through the glass transition and are relatively stable.

Amorphous semiconductors which have been studied most extensively are tetrahedrally-bonded materials, such as a-Si, and chalcogenide glasses, both elemental or compound. While in the latter the connectivity of chemical bonds is either of a one-dimensional nature, chains and rings as in the case of elemental chalcogenide glasses, or of a two-dimensional nature, layers or certain cross linked networks, the four-fold coordination in the former yields three-dimensionally interconnected networks. It is not difficult to understand that the parameters describing the atomic correlations are more constrained in three-dimensional random connectivities than in lower-dimen-sional networks. This is one of the reasons why the structural models are easier to construct for tetrahedrally-bonded amorphous semiconductors and why more theoretical analysis has been carried out.

3. Short-range order, long-range disorder and gaps

The most remarkable difference between crystals and amorphous materials lies in the fact that the long-range periodic order characteristic of the former no longer exists in the latter while the short-range order does not differ very much. The absence of strict long-range order is common to all dis-ordered systems. The one-electron theory of crystals has been developed on the basis of Bloch's theorem in which long-range order, i.e. periodicity is the key factor. One of the most remarkable outcomes of the electron theory of crystals is the fact that it gives a microscopic explanation of the dif-ference among metals, semiconductors and insulators in terms of the concepts of energy bands and gaps. Perhaps, because of this history, there existed an implicit, vague idea that the bands and gaps arose from long-range order. The inadequacy of this idea might be readily realized by noting that window glass is transparent! For a long time, we have known that the window glass is a typical noncrystalline material, and yet the fact that the photon energy of visible light cannot excite electrons implies the existence of a gap larger than this energy.

The assertion due to Ioffe and Regel[8], that most of the important phys-ical quantities are determined mainly by short-range order, was one of the early warnings to adherents to the Bloch mystique. Unexpectedly small changes of resistivities on melting is one example of the observations which lend support to the assertion[9].

In some amorphous semiconductors as well, experiment has suggested a quite well-defined gap, of the same magnitude as found in a corresponding crystal, when the defects are successfully removed leaving the system pre-dominantly topologically disordered. In the conduction band tail of a-Si:H for instance, the density of states changes by an order of magnitude over only 0.1 eV [10]. The minimum density of states in the gap of a-Si:FH is 10^5 times lower than its value at the conduction band mobility edge [11].

These observations are not only academically interesting but are also important in the context of the technological need for narrow band tails and a deep minimum in the density of states. Thus, we are led to ask why the band tails are so narrow, how narrow they can get, and how deep a minimum can be achieved.

In what follows, we focus our attention on this problem and present the first stages of an attempt to answer these questions in a study of ideal a-Si (or a-Ge).

4. Spectral bounds and band tails for alloys

For later convenience, let us now show what has been learned about the spectral bounds and band tails of a disordered binary (AB) alloy. In the one-dimensional case, the famous Saxon-Hutner theorem [12] was originally stated and later proved in the form: *any spectral region that is a spectral gap for both a pure A-type chain and a pure B-type chain is also a gap for any mixed lattice of A- and B-type atoms.* For higher dimensions, we have a general theorem for the bounds of the spectrum of any tight-binding (TB) model which also includes one-dimension [13-15].

Let us study a TB Hamiltonian operator given by

$$H = \sum_i |i> \varepsilon_i <j| + \sum_{i \neq j} |i> V_{ij} <j| \quad , \tag{4.1}$$

where the sites $\{i\}$ form a regular lattice, the $|i>$ are the orthonormal Wannier functions associated with sites $\{i\}$, and V_{ij} is nonzero only when i and j are nearest neighbors. Suppose we have diagonal disorder alone so that $V_{ij} \equiv V =$ constant irrespective of the kinds of atoms on i and j, while ε_i is equal either to ε_A or to ε_B according as the site i is occupied by an A or B atom, respectively.

An eigenfunction of H defined by $H|\psi> = E|\psi>$ is expressed as $|\psi> = \Sigma_i \, a_i |i>$ for each eigenvalue E. The Schrödinger equation for the Hamiltonian (4.1) and this Ψ is

$$(E - \varepsilon_i)a_i = \sum_{j \neq i} V_{ij} \, a_j \quad ; \quad i = 1, \cdots, N \, , \tag{4.2}$$

where N is the total number of lattice sites. From (4.2), we get the inequality

$$|E - \varepsilon_i| \leq \sum_{j \neq i} |V_{ij}| \{|a_j|/|a_i|\} \quad . \tag{4.3}$$

Now suppose that i labels the site where this eigenfunction has maximum amplitude; then the ratio $\{|a_j|/|a_i|\}$ is always less than unity, so that

$$|E - \varepsilon_i| \leq \sum_{j \neq i} |V_{ij}| = z|V| \equiv B \, , \tag{4.4}$$

where z is the number of nearest neighbor atoms (the coordination number). For every eigenvalue E, there exists at least one site, i, for which this inequality holds. Equation (4.4), generally known as the Hadamard-Gerschgorin theorem of matrix algebra, gives rise to the following statement about the spectral bounds: the spectrum must lie wholly within the region covered by two bands, each of the 'perfect' width (2B), centered on the local levels ε_A and ε_B of the two constituent atoms. This shows that a gap must open up when $|\varepsilon_A - \varepsilon_B| > 2B$. The Saxon-Hutner theorem is obviously included in (4.4).

In this way, it has been proved that the gap is not a consequence of long-range order.

Let us now turn to the problem concerning the behavior of the band tails of the AB alloy. Lifshitz [16] deduced a theorem from the local density principle. We consider the case where $\varepsilon_A < \varepsilon_B$. Suppose that ε_0^A is the lower edge of the lower band A. This bound would be reached in an infinite crystal of type A atoms. The addition of each subsequent B atom can only have the effect of increasing each eigenvalue. To get an eigenvalue E near ε_0^A, we must look for a sufficiently large region that has been left free of B atoms by statistical fluctuations of local concentration. Within this region of linear dimension L, there exist standing waves whose smallest wave number is $k \sim \pi/L$, which is related to the lowest level by

$$ E - \varepsilon_0^A = \frac{\hbar^2 k^2}{2m} = \frac{\hbar^2}{2m} \left(\frac{\pi}{L}\right)^2 . \tag{4.5} $$

In a lattice of d spacial dimension, such a region would contain

$$ n \sim (L/a)^d \tag{4.6} $$

sites, where a is a length of the order of the lattice spacing. When we suppose the concentration of A atoms to be c_A, the probability of finding such a region (in the complete absence of correlation) is expressed as

$$ (c_A)^n \sim \exp[(L/a)^d \ln c_A] , \tag{4.7} $$

which we interpret as the probability of finding a state of the chosen energy E. Then, we get an estimate of the density of states in the tail as

$$ N(E) \propto \exp[-C(E - \varepsilon_0^A)^{-d/2}] , \tag{4.8} $$

where C is a numerical factor of the order of unity when the energy is scaled by the band width.

5. Short-range correlation in tetrahedrally-bonded materials

We classified disordered systems in §2 where we categorize amorphous semiconductors as having primarily topological disorder. Our classification is based on the experimental observation of strong short-range correlation in amorphous solids.

Let us take a-Si as an example. The position and width of the first peak of the radial distribution function (RDF) gives information about the bond length r and its fluctuation Δr respectively; $\Delta r/r$ is found to be less than 1%. The area under the first peak is identified as the coordination number, which is almost four. The position and width of the second peak together with those of the first peak give an estimate for the bond angle θ and its fluctuation $\Delta\theta$; it has been found that $\Delta\theta/\theta \sim 8\%$ [17].

In order to obtain further information about short-range correlation, higher-order atomic-distribution functions are required. Since distributions higher than the pair distribution are experimentally not available so far, several structural models have been constructed either manually or using computers by keeping the conditions $\Delta r \lesssim 0.01$ and $\Delta\theta/\theta \sim 0.08$ [18-21]. From the analysis of the continuous random network thus constructed, the existence of substantial _dihedral-angle fluctuation_ and _odd-membered rings_ is inferred.

All these types of disorder are not independent of one another. For instance, when all bond lengths, bond angles and dihedral angles are completely regular, then odd-membered rings cannot be present. Only six-membered rings occur. Therefore, it is somewhat artificial to treat them individually as if they were independent, but it is nevertheless convenient for the purpose of establishing a direct connection between causes on the atomic scale and their effects on macroscopic observations.

With this in mind, we concern ourselves with disorder in bond lengths, bond angles, dihedral angles and rings. The former three types of disorder are hereafter referred to as quantitative disorder while the ring disorder is regarded as purely topological. It is very difficult to determine the dihedral-angle distribution experimentally. The existence of odd-membered rings has so far not been directly ascertained by experiment. Nevertheless the possible importance of dihedral-angle disorder and ring disorder has been noticed for some time.

Before we study previous attempts to examine the effects of odd-membered rings, it is appropriate to mention the experimental results for the energy spectrum. The density of states (DOS) of the valence band of Ge determined from XPS is shown in Fig.3(a)[23] for both crystalline and amorphous states. As is obvious from the figure, the gross features of the DOS of c-Ge are similar to those of a-Ge. A striking aspect is that both have almost the same band width. Two significant differences between c-Ge and a-Ge are first that the two lower peaks II and III for c-Ge coalesce into a single broad peak and second that the topmost peak I remains essentially unchanged but slightly skewed to the right. The DOS of Si shows analogous features (Fig.3(b))[23].

Several attempts have been made to reproduce theoretically the observed DOS and extract its essence in amorphous structures. One assertion is that polymorphs of Ge with symmetry lower than diamond-cubic structure fall somewhere between the diamond structure of high symmetry and the amorphous state of no symmetry [24]. The valence-band DOS of ST-12, which includes five-membered rings, shows a fair degree of resemblance to the experimental data with two distinct peaks II and III, while, in the calculated DOS of other tetrahedrally bonded structures with even-membered rings alone, the peaks II and III persist. The importance of odd-membered rings for the disappearance of the dip between peaks II and III has also been supported by other work [25-26].

The idea of ring topology of this kind is certainly interesting and its existence may really play an important role. However, it has also been reported that other structural models with only even-membered rings to have a DOS with peaks II and III merged [27]. A point to be made, therefore, is

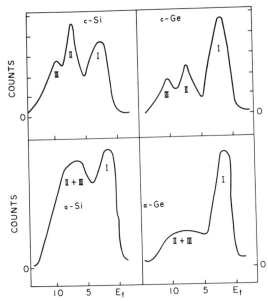

Fig.3 XPS data of the valence band DOS of (a) c-Ge and a-Ge; and
(b) c-Si and a-Si (after Ley et al. 1972 [23]).

that a structural model and a macroscopic physical property are not necessarily related by a one-to-one correspondence. There might be more than
one possible structural model which yields particular observable properties.
It is important, therefore, to see what can be deduced from a given structural model as well as what cannot; by elimination, it may be possible to
obtain more reliable information.

6.   Analysis of the Weaire-Thorpe model

We study in this section the Weaire-Thorpe model Hamiltonian and see how far
we can go with it.

6.1 The Weaire-Thorpe model

The model Hamiltonian operator due to Weaire and Thorpe [28-30] is described
in the tight-binding form:

$$H = \sum_i \sum_{\mu=\mu'} |i\mu> V_1 < i\mu'|, + \sum_i \sum_\mu |i\mu> V_2 < i_\mu\mu| \quad , \qquad (6.1)$$

where $|i\mu>$ denotes a directed hybridized orbital of the sp^3 type associated
with site i and bond μ; i_μ denotes the nearest neighbor of i connected by
bond μ. Matrix elements of the Hamiltonian are assumed to be nonvanishing
only between orbitals associated with the same atom (V_1) or the same bond
(V_2), (see Fig.4). The orbitals $\{|i\mu>\}$ are regarded as forming a complete,

126

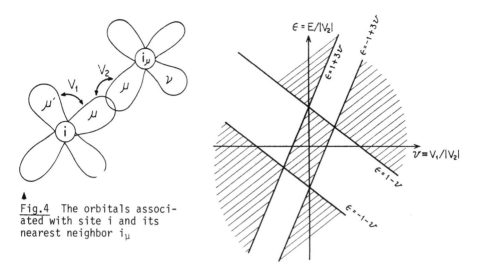

Fig.5 Spectral bounds. $\varepsilon = E/|V_2|$, $v = V_1/|V_2|$. Shaded regions are forbidden, unshaded allowed. Thus it is readily seen that there does exist a gap for any arbitrary set of values for V_1 and V_2 (after Weaire 1971 [31]).

orthonormal basis. It is usually more convenient to add a self interaction to the V_1 term in H, shifting all energies by V_1, so that (6.1) is written as

$$H = \sum_i \sum_{\mu,\mu'} |i\mu> V_1 <i\mu'| + \sum_i \sum_\mu |i\mu> V_2 <i_\mu\mu| \quad . \qquad (6.1a)$$

The bond-angle disorder is reflected primarily in the matrix elements $\{V_1\}$, while the bond-length disorder appears primarily in $\{V_2\}$. Information concerning the connectivity of atoms manifests itself in summations over i, i_μ and μ,μ'. Actually, we can use (6.1) for any kind of connected structure by choosing the summations in an appropriate manner.

Weaire and Thorpe have discussed the consequences of topological disorder for amorphous semiconductors, in particular Si and Ge in a fully interconnected continuous random network (CRN) of four-fold coordinated atoms. By taking V_1 and V_2 constant all over the system, bounds to the sp^3-derived valence and conduction bands and an energy gap between them have been demonstrated to exist for otherwise arbitrary structures. (See Fig.5)[31]. They also have shown that the DOS for this model is obtained once we evaluate the DOS of a simple s-band Hamiltonian for the same topological disorder. Recent experiments seem to be suggesting that ideal, pure a-Si may exist in concept only; nevertheless it is still important and instructive to examine this simple yet non-trivial model in order to learn some essential aspects of amorphous silicon. A more realistic but more complicated model could not possibly be solved unless we have a good knowledge of the simpler model.

6.2 Mapping of energy levels and eigenfunctions

As has been mentioned, the work by Weaire and Thorpe is useful because the existence of the gap has been proved for topological disorder as well as because they have shown how to obtain the energy levels of this model from the knowledge of those for a single s-band. We have extended their work to show that the Weaire-Thorpe model can be mapped onto the nearest-neighbor, tight-binding s-band model not only as regards energy levels but also as regards the eigenfunctions. This is important since, through this mapping, we can practically solve the Weaire-Thorpe model by solving a much simpler single s-band model. Actually, the s-band model is the simplest of all tight-binding models, but the primary effects of topological disorder are already manifested by it.

Let us write an eigenfunction specified by λ as

$$|\Psi_\lambda> = \sum_{i,\mu} a^\lambda_{i\mu} |i\mu> \quad , \quad (\lambda = 1, 2, \cdots\cdots, 4N) \quad , \tag{6.2a}$$

where the normalization is such that

$$\sum_{i,\mu} |a_{i\mu}|^2 = 1 \quad . \tag{6.2b}$$

In what follows, we drop the suffix λ for simplification. The formalism holds for each eigenvalue $E_\lambda = E$, which is given by

$$E = <\Psi|H|\Psi>$$

$$= V_1 \sum_i |\sum_\mu a_{i,\mu}|^2 + V_2 \sum_{i,\mu} a^*_{i,\mu} a_{i_\mu,\mu}$$

$$\equiv V_1 x + V_2 y \quad , \tag{6.3}$$

where

$$x \equiv \sum_i |\sum_\mu a_{i,\mu}|^2 \quad , \tag{6.4}$$

$$y \equiv \sum_{i,\mu} a^*_{i,\mu} a_{i_\mu,\mu} \quad . \tag{6.5}$$

Using a Schwartz inequality, it is easy to show that [32,33]

$$0 \leq x \leq 4 \quad , \tag{6.6a}$$

$$-1 \leq y \leq +1 \quad . \tag{6.6b}$$

Now, let us see what states the bounds of x and y correspond to:
(1) The bound $x = 0$ is realized when $\sum_\mu a_{i\mu} = 0$ for each i; this means that the eigenfunction is p-like at each site i.
(2) The bound $x = 4$ is realized when, for a given i, $a_{i\mu}$ all have the same value for different μ; this means that the eigenfunction is s-like at each site.
(3) The bound $y = -1$ is realized when $a^*_{i\mu} a_{i_\mu\mu} = -1/4N$ on each bond, corresponding to an antibonding state, while

(4) The bound $y = +1$ is realized when $a^*_{i_\mu} a_{i_\mu,\mu} = 1/4N$ on each bond, corresponding to a bonding state.

The bounds of x and y define the bounds of eigenvalues E. It must be noted, however, that x and y are dependent on each other, and cannot be varied independently. In order to find out how x and y are related, let us study the equation of motion for $a_{i,\mu}$

$$E\ a_{i,\mu} = V_1 D_i + V_2\ a_{i_\mu,\mu} \quad , \tag{6.7}$$

where

$$D_i \equiv \sum_\mu a_{i,\mu} \quad . \tag{6.8}$$

Using (6.7) and a similar equation for $a_{i_\mu,\mu}$ we have

$$[E^2 - V_2^2] a_{i,\mu} = V_1 [E\ D_i + V_2 D_{i_\mu}] \quad . \tag{6.9}$$

The summation over μ yields

$$f(E) D_i - V_2 \sum_{\mu=1}^{4} D_{i_\mu} = 0 \quad , \tag{6.10}$$

where

$$f(E) = [E^2 - V_2^2 - 4V_1\ E]/V_1 \tag{6.11}$$

and i_μ denotes the neighbors of i.

Now it is readily seen that (6.10) is equivalent to the equation of motion for the s-band Hamiltonian

$$H_s = \sum_{(i,j)=n.n.} |i> V_2 <j| \quad , \tag{6.12}$$

if we identify f(E) as the eigenvalue ε of (6.12) and D_i as the amplitude at i of the eigenfunction of (6.12). This completes the proof of the mapping both with respect to the spectrum and to the eigenstates. In practice, the DOS per atom, N(E), of the Weaire-Thorpe model can be evaluated if the dispersion ε_ν of (6.12) is known; ie, the DOS per atom, $n(\varepsilon)$, of (6.12) is expressed in terms of ε_ν as

$$n(\varepsilon) = N^{-1} \sum_{\nu=1}^{N} \delta(\varepsilon - \varepsilon_\nu)$$

$$= -(N\pi)^{-1}\ \mathrm{Im} \sum_{\nu=1}^{N} \frac{1}{\varepsilon^+ - \varepsilon_\nu} \quad . \tag{6.13}$$

Then, the DOS of (6.1) is derived as follows;

$$N(E) = N^{-1} \sum_{\nu=1}^{N} \delta(f(E) - \varepsilon_\nu)$$

$$= -(N\pi)^{-1} \text{ Im } \sum_{\nu=1}^{N} \frac{2(E - 2V_1)}{(E^+ - 2V_1)^2 - (4V_1^2 + V_2^2) - V_1 V_2 \tilde{\varepsilon}_\nu} \quad , \quad (6.14)$$

where $\tilde{\varepsilon}_\nu = \varepsilon_\nu/V_2$. The total density $\int_{-\infty}^{\infty} N(E)dE$ is 2 per atom.(Fig.6.)

When all $\{D_i\}$ are zero simultaneously in (6.10), we have from (6.9)

$$E^2 - V_2^2 = 0 , \qquad (6.15a)$$

or

$$E = \pm V_2 . \qquad (6.15b)$$

These two levels appear as two delta-function-like levels in the DOS. The fact that $D_i=0$ at each site means that the function is p-like on each site, and therefore these two levels are p-like.

From (6.9) and (6.10), we can derive

$$[E^2 - V_2^2][2xV_1(E - 2V_1) - \{E^2 - V_2^2\}] = 0 \qquad (6.16)$$

$$x-2 = \frac{1}{2V_1}[(E - 2V_1) - \frac{V_2^2 - 4V_1^2}{(E - 2V_1)}] \qquad (6.17)$$

$$y = \frac{1}{2V_2}[(E - 2V_1) + \frac{V_2^2 - 4V_1^2}{(E - 2V_1)}] , \qquad (6.18)$$

and

$$-V_1^2(x-2)^2 + V_2^2 y^2 = V_2^2 - 4V_1^2 \quad , \qquad (6.19)$$

for a-Si and a-Ge, the estimated values for V_1 and V_2 satisfy the relation $V_2^2 - 4V_1^2 > 0$.

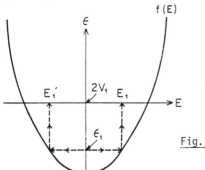

Fig. 6   The mapping from ε to E through

$$\varepsilon \equiv f(E) = [(E-2V_1)^2 - \{4V_1^2 + V_2^2\}]/V_1$$

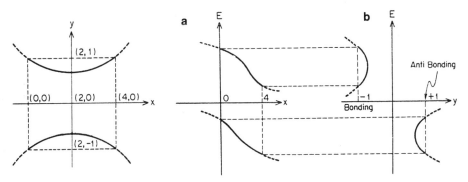

Fig. 7 The correlation between x and y; note $0 \leq x \leq 4$ and $-1 \leq y \leq 1$

Fig. 8 (a) The x-E curve; and (b) the y-E curve

The correlation between x and y is given in Fig.7, from which we can see that, for a given value of x ($0 < x < 4$), there exist two values for y. This implies that each state in the bonding band has a corresponding state in the anti-bonding band. The change of x as a function of energy E is illustrated in Fig.8(a) while the y vs E relation is shown in Fig.8(b). The existence of a gap is obviously guaranteed.

6.3 A single s-band

With a view to obtaining explicit results for the Weaire-Thorpe model through the mapping explained in the preceding subsection, we illustrate in this subsection the results of our investigations of the influence of topological disorder on tight-binding energy bands by reporting them for the band shape and eigenstates of the s-band.

The Hamiltonian is

$$H_s = \sum_{i,\mu} |i> V <i_\mu| \quad , \tag{6.20}$$

which is the same as (6.12) with $V = V_2$. Here, we suppose $V < 0$. The energy of an eigenstate with amplitude a_i on the ith site is

$$\varepsilon = z V y \quad , \tag{6.21}$$

$$zy = \sum_{i,\mu} a_i^* a_{i_\mu} \quad . \tag{6.22}$$

A Schwartz inequality obeyed by y ($-1 \leq y \leq 1$) imposes the bounds zV, $-zV$ on the spectrum.

At the lower bound, the eigenfunction is bonding, $a_i = N^{-1/2}$, where N is the number of sites. The spectrum can be established near zV by a perturbation procedure. Consider a set of N plane waves

$$x_{\vec{k}} = N^{-1/2} \exp[i\vec{k} \cdot \vec{r}] \quad ,$$

131

where $\vec{r}_i$ is the position of site i, with the wave-vectors $\vec{k}$ chosen so as to satisfy some convenient boundary conditions. We confine ourselves to what we term normal structures. For these the most probable value of the overlap matrix $(x_{\vec{k}}, x_{\vec{k}}') = N^{-1/2} \Sigma_i \exp[i\vec{Q} \cdot \vec{r}_i] \equiv A_Q$, $\vec{Q} = \vec{k} - \vec{k}'$, is zero in an ensemble of equivalent structures. The mean square value

$$<|A_{\vec{Q}}|^2> = N^{-1} < \Sigma_{i,j} \exp[i\vec{Q} \cdot (\vec{r}_i - \vec{r}_j)]> = N^{-1} S(Q) \quad ,$$

where $S(\vec{Q})$, the structure factor, is $O(1)$. Thus, as $N \to \infty$, the $A_{\vec{Q}}$ become sharply distributed about zero, and the set $x_{\vec{k}}$ can be regarded as orthonormal. We use the $x_{\vec{k}}$ as a basis for our perturbation procedure.

The eigenfunctions of H can be written in the form

$$\Psi_{\vec{k},i} = N^{-1/2} e^{i(\vec{k} \cdot \vec{r}_i + \phi_i)}$$

with ϕ_i an arbitrary complex function of i. Because ϕ_i becomes a constant which can be taken as zero when $k = 0$, we can try an expansion of ϕ_i in powers of k for small $k \equiv |k|$ and test for convergence to lowest order in k. The result is

$$\varepsilon = zV + \hbar^2 k^2 / 2m^* \tag{6.23}$$

$$\Phi_i = V(G(zV)\vec{k} \cdot \vec{\rho})_i \tag{6.24a}$$

$$(m^*)^{-1} = \frac{1}{3} \frac{|V|}{\hbar^2 N} \Sigma' |\vec{\rho}_\alpha|^2 [1 - \frac{2V}{zV - \varepsilon_\alpha}] < m_0^{-1} = \frac{1}{3} \frac{|V|}{\hbar^2 N} \Sigma' |\vec{\rho}_\alpha|^2 \quad . \tag{6.24b}$$

In (6.24) $G(zV)$ is the Green's function corresponding to (6.20) but with the bonding state projected out and $\vec{\rho}_i = \Sigma_\mu (\vec{r}_{i\mu} - \vec{r}_i)$. In (6.25) ε_α is an eigenvalue of (6.20) and the sum excludes zV. Similarly $\vec{\rho}_\alpha = \Sigma_i \vec{\rho}_i S_{i\alpha}$ with $S_{i\alpha}$ the transformation diagonalizing (6.20). The perturbation expansion in powers of $\vec{k}$ converges in probability for sufficiently small $\vec{k}$ provided only that $\vec{\rho}_i$, which has been assumed to have mean value zero, has bounded moments. We take this to be true for normal structures. In a structure such as a crystal for which all $\vec{\rho}_i = 0$, ϕ_i in (6.24a) would go to zero, and m* would go to m_0. In the general case, $N^{-1/2} \exp[i\vec{k} \cdot \vec{r}_i]$ would not be an eigenfunction but simply a trial function, giving $m^* > m_0$ as in (6.25). The $\Sigma'_\alpha |\vec{\rho}_\alpha|^2$ in $(m_0)^{-1}$ in (6.24b) can be rewritten as $\Sigma \rho_i^2$. Eq.(6.23) implies a DOS $n(E) \propto (E - zV)^{1/2}$ near the bonding bound. We call such a bound a normal band edge.

The moments of the entire band

$$M_n = \int \varepsilon^n n(\varepsilon) d\varepsilon \tag{6.25}$$

are given by

$$M_0 = 1; \quad M_1 = 0; \quad M_2 = zV^2; \quad M_n = W_n V^n, \quad n > 2 \quad . \tag{6.26}$$

In (6.26), W_n is the average number of distinct, closed, n-sided polygons which are constructed by walks on the network starting from a given site and taking steps only between nearest neighbors. If there are no polygons with n odd, i.e. the structure is even (bichromatic), all odd moments vanish, and $n(E)$ is even about $E = 0$. The bound $-zV$ is also a normal band edge corresponding to the antibonding state, with $y = -1$ and $a_j = \pm N^{-1/2}$ changing sign across each nearest neighbor bond in the network.

The general n-sided polygon can be decomposed into rings, which are polygons derived from non-self-intersecting walks on the network. The presence of odd rings of order m and greater implies the presence of odd polygons of all orders $>m$. Real structures, those of amorphous semiconductors, can be supposed always to contain odd rings. The symmetry in $n(E)$ about $E = 0$ is then lost. More important, the character of the states near the antibonding limit changes radically. An antibonding orbital with $a_j = \pm N^{-1/2}$ and $y = -1$ which changes sign across each bond is incompatible with the structure. The wavefunction would change sign an odd number of times in walking round an odd ring and could not be single valued. An extended state with $a_j = \pm N^{-1/2}$ thus cannot change sign across every bond. Let n be the minimum possible number of frustrated bonds, those across which the sign does not change. The corresponding state is not an eigenfunction of (6.20) because the random distribution of frustrated bonds would lead to amplitude as well as phase variations. However, the expectation value of the energy,

$$\varepsilon_{op} = - (z - 4n/N)V \tag{6.27}$$

is a kind of optical potential. States near or above E_{op} in energy can be expected to be profoundly affected by the topological disorder and, in particular, by the ring statistics.

Because the rings and therefore the frustrated bonds are randomly distributed, we can imagine a fluctuating density of frustrated bonds and, in particular, regions which are free of them and therefore even. The problem becomes exactly analogous to that considered by Lifshitz for alloys if the frustrated bonds are completely randomly distributed. The antibonding bound is reached with zero probability, and the DOS is given by

$$n(\varepsilon) = n_0 \exp[-(\varepsilon a/-zV-\varepsilon)^{3/2}] \tag{6.28}$$

where εa is a constant. In the more general case, where the frustrated bonds are correlated, the overall shape of the tail is similar but the energy dependence differs from (6.28). We term a bound which is realized with zero probability a band limit. The antibonding bound is always a limit in disordered structures containing odd-rings and, when the frustrated bonds are uncorrelated, it is a Lifshitz limit (6.28).

The antibonding states in the tail (6.28) are localized, being confined to locally even regions. On the other hand, the bonding states near zV are extended. There must be at least one mobility edge within the band, and, presuming on the simplicity of the band model, we suppose there to be only one. Because we are dealing with states in the upper half of a doubly-bounded continuum, the variational principle tells us that relaxing the amplitude constraint on our extended state $a_j = \pm N^{-1/2}$ would raise its energy. Thus the mobility edge ε_c is above ε_{op}. There is no reason to expect localized states and mobility edges for an even structure.

We learn from the above that, in such a mixed structure, the density of states is transferred downward from the vicinity of the antibonding bound. From the vanishing of the first moment, we infer that the transferred density of states must remain above $\varepsilon = 0$, indicating a possible peak above $\varepsilon = 0$ for sufficient disorder.

A schematic representation of the density of states of the topologically disordered s-band and of those other features which we have elicited for even and asymmetric structures is shown in the Fig.9. Perhaps the most striking results are that the bonding bound is a normal edge and the antibonding bound a limit for an asymmetric structure. These results, or related ones, can be expected to obtain for more complex band models as well.

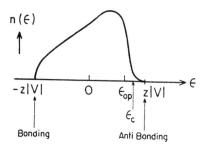

Fig.9 The density of states (DOS) of a topologically disordered nearest-neighbor s-band.

6.4 Mapping of the energy bands

Using the results of preceding two subsections (6.2 and 6.3), the DOS of the Weaire-Thorpe model is evaluated as follows: As shown by Eq.(6.15b), Eq.(6.9) has 2N independent pure p solutions with $D_i = 0$ and $E = \pm V_2$, corresponding to x = 0 and y = ± 1, respectively. There is one bonding p-state and one anti-bonding p-state per atom associated with $\delta(E \pm V_2)$ in $N(E)$. Moreover, as we have seen in §6.2, Eq.(6.14) gives an explicit mapping to the DOS for the remaining 2N levels in the Weaire-Thorpe model from the DOS for the s-band in the same structure. Equations from (6.16) to (6.19) demonstrate that the DOS is symmetric under the transformation $(E - 2V_1) \rightarrow -(E - 2V_1)$, which simultaneously changes x to 4-x, interchanging the s- and p-parts of the wavefunctions, and y to -y, interchanging the bonding and antibonding parts of the wavefunctions. The actual mapping of the DOS through f(E) is illustrated in Fig.10 [34].

The two p states per atom associated with the δ-function peaks are in fact π states. The δ-function at the top of the valence band corresponds to peak I, which, in the Weaire-Thorpe model is pure π. The rest of the valence and conduction bands are pure σ states so that the mapping from the s-band to the Weaire-Thorpe model is in fact a mapping from s to σ states. Peaks II and III persist in the Weaire-Thorpe model for the diamond structure but merge for the topologically disordered case, according to the mapping. Odd rings emphasize the merged peak and the dip between it and peak I.

7. Dihedral-angle disorder

Through the Weaire-Thorpe model, despite its simplicity, we have learned some essential aspects of the effects due to topological disorder. In addition, the model is extremely advantageous because the mapping to the simplest s-band model is possible. Nevertheless, it is necessary to improve the model to encompass the complexity of realistic systems. In particular, elimination of the delta-function-type degeneracy inherent in the DOS of the model is vital. Besides, there is no way of studying the effects of dihedral angle disorder (DAD) within this model.

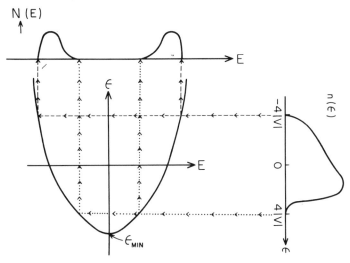

Fig. 10 The mapping of the DOS through the function

$$\varepsilon \equiv f(E) = [(E-2V_1)^2 - \{4V_1^2 + V_2^2\}]/V_1 .$$

There are several possible ways to extend the original Weaire-Thorpe model. One direct way is to introduce quantitative disorder in V_1 and V_2, each respectively corresponding to fluctuations in θ and r. When disorder in V_2 is included, the delta-function type singularity in the DOS is removed. This actually is not an essential improvement of the Weaire-Thorpe model because the delta-function degeneracy cannot be eliminated in the case of a crystal for which no such degeneracy is found in actuality. Besides, it is expected that the energy levels are not much shifted by the inclusion of fluctuations in V_1 and V_2 as we shall estimate below for the edges. Moreover, nothing typical of the amorphous structure is learned by studying quantitative disorder in V_1 and V_2.

The effects of $\Delta\theta$ and Δr may be estimated as follows. Since there are very strong short-range spatial correlations present in the random covalent network, a tight-binding model Hamiltonian is most convenient for the study of the electronic structure. Following Slater and Koster [35], we obtained tight-binding parameters up through second neighbors (13 terms) by fitting to 0.1 eV the levels at Γ, X, and L calculated by Chelikowsky and Cohen [36] for Si in the diamond structure. No attempt at optimization was made, but the resulting energy bands agree satisfactorily with the pseudopotential calculations up to 3.5 eV into the conduction band, which is sufficient for our purposes. Fig.11 shows the fit, and Table 2 the parameters [1,5]. To transfer these parameters to a-Si, it is necessary to know how they depend

Table 2 Thirteen parameters used for the fit in Fig.11

Matrix Element	$E_{ss}(0)$	$E_{xx}(0)$	$E_{ss}(\frac{1}{2},\frac{1}{2},\frac{1}{2})$	$E_{xx}(\frac{1}{2},\frac{1}{2},\frac{1}{2})$	$E_{xy}(\frac{1}{2},\frac{1}{2},\frac{1}{2})$	$E_{sx}(\frac{1}{2},\frac{1}{2},\frac{1}{2})$	$E_{xx}(011)$
Value in Ex	1.0	-5.4	-2.05	0.43	1.20	1.13	-0.25

Matrix Element	$E_{ss}(110)$	$E_{xx}(110)$	$E_{xy}(110)$	$E_{sx}(110)$	$E_{xy}(110)$	$E_{sx}(001)$
Value in Ex	0.11	0.21	0.07	0.03	0	0.03

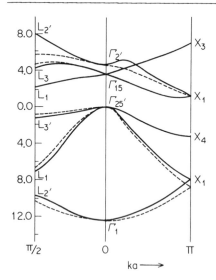

Fig. 11 Band structure of crystalline silicon; —— Ref.36, --- our fit (Cohen, Singh and Yonezawa, [1])

on bond lengths and angles. This we ascertain by fitting the deformation potentials of the crystal by our tight-binding model, including spin-orbit coupling. A detailed analysis of the band-structure fit shows that it is sufficient to make the molecular approximation for those parameters which are nonvanishing with full cubic symmetry. The resulting deformation potentials are shown in Table 3 [1,5].

Several structural models of a-Si have been constructed which are consistent with its radial distribution function and density. From these models we have estimated that the fluctuations in bond length r about its crystalline value are Gaussian with rms deviation $\Delta r/r = 0.015$. The bond angle fluctuates about the tetrahedral angle 109.3° with an rms deviation of $\Delta\theta = 10°$. The local structure variations cause local variations ΔE in the matrix elements and therefore quantitative disorder in the Hamiltonian, which we find relatively small. To gain a clearer idea of the importance of the quantitative disorder, we have estimated the shifts in $\Gamma_{25'}$ (Γ_8-) and X_1 in the valence band of the crystal caused by changes in the crystalline matrix elements equal to the rms variation of those same parameters in the amorphous materials. Table 4 shows the shifts caused separately by the bond-length and bond-angle variations.

Table 3 Significant deformation potentials determined from modulation spectroscopy
results for silicon. $\Delta E(\nu)_i = D(\nu)_i \Delta R_i$.

Deformation Potential	$\frac{\sqrt{3}a}{2}\, D(pp\sigma)_1$	$\frac{\sqrt{3}a}{2}\, D(pp\pi)_1$	$\frac{\sqrt{3}a}{2}\, D(pp\sigma)_1$	$\frac{\sqrt{3}a}{2}\, D(sp\sigma)$	$\sqrt{2}a\, D(pp\sigma)_2$	$\sqrt{2}a\, D(pp\pi)_2$
Value in eV	-1.9	0.9	2.6	1.5	1.6	-0.6

Table 4 Approximate changes produced in $\Gamma_{25'}$ and X_1 due to bond length (ΔR) and
bond angle ($\Delta\theta$)

Structural change	Change in $\Gamma_{25'}$ (eV)	Change in X_1 (eV)
ΔR	0.02	0.025
$\Delta\theta$	0.2	0.1

To examine the effects of dihedral-angle and topological disorder, we
keep all of the nearest neighbor interactions. We summarize in the rest of
this section the results of Singh [5] for that model. Using an sp^3 bonding
orbital basis, there are six interactions, V_1, V_2, $\cdots$, V_6, as shown in
Fig.12. The Hamiltonian may be written as

$$H = \sum_{i,\mu,\mu'} |i\mu> V_1 <i\mu'| + \sum_{i,\mu} |i\mu> V_2 <i_\mu\mu|$$

$$+ \sum_{i,\mu\neq\mu} |i\mu> V_3 <i_\mu\nu| + \sum_{i,\mu\neq\mu,\alpha} |i\mu'> V_\alpha^i\, i\mu <i_\mu\nu_\alpha| \quad .$$

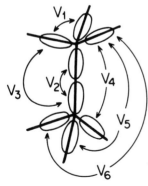

Fig. 12 Definition of six interactions $V_1,\ldots,\ V_6$
between orbitals associated with nearest-neighbor atoms

137

The index α takes on the values 4, 5 and 6. The interaction V_α depends on the dihedral angle of rotation of the bond ν_α of site $i\mu$ around the i-i_μ bond relative to the bond μ' of site $i\mu$ requiring the superscript i, i_μ. We use a convention in which V_4 is the interaction with the furthest bond, V_5 with the intermediate bond, and V_6 with the closest so that $|V_6| \geq |V_5| > |V_4|$; 120° periodicity in the dihedral-angle Φ must be maintained. V_4, V_5, and V_6 are special cases of a general interaction $V(\Phi)$.

In the case of a diamond cubic crystal, the dihedral angles are characterized by the staggered configuration; thus Φ_4=180° and Φ_5=Φ_6=60° and we have $V_5 = V_6$. For a crystal where all these interactions V_1 to V_6 are constant, it is easily shown that the delta-function type degeneracy is not removed unless $V_4 \neq V_5$ (=V_6). For illustration, the effects of these interactions on the band structure are shown in Figs.13(a) and (b), where Fig. 13(a) corresponds to the Weaire-Thorpe model and Fig.13(b) to the case where V_1, V_2 and V_3 are constant and $V_4 \neq V_5$ (=V_6). Fig.13(c) is the result of an empirical pseudopotential calculation.(After [37]).

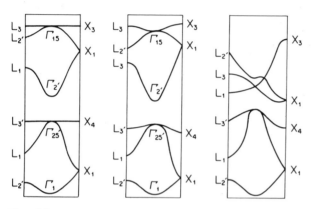

Fig. 13

(a) The band structure of the Weaire-Thorpe model for a diamond-cubic crystal with V_1 = -2.5 eV and V_2 = -7.53 eV.
(b) The band structure of the model Hamiltonian Eq.(7.1) for a diamond-cubic crystal with V_1 = -2.5 eV, V_2 = -7.53 eV, V_3 = 0, V_4 = 0.26 eV, $V_5 = V_6$ = 0.
(c) The result of the pseudopotential calculation.
(After [37]).

In disordered systems, the interactions V_1, V_2, and V_3 are invariant under dihedral-angle rotations; the influence of dihedral-angle disorder cannot be examined without including V_4, V_5 and V_6. Numerical values for all of the V's are determined by a tight-binding fit including second-neighbor interactions to the self-consistent pseudopotential calculations of the energy bands of crystalline (c-) Si. From these values we find the variation of $V(\Phi)$[5],

$$V(\Phi) = 0.3 - 1.0 \cos^2 \Phi/2 \quad . \tag{7.2}$$

There is a 1 eV variation in $V(\Phi)$, and accordingly dihedral-angle disorder must play an important role in the electronic structure of a-Si. The bounds of energies can be estimated by checking the bounds of each term. For the wave function (6.2), the energy is [5]

$$E = V_1 x + V_2 y + V_3 w + \sum_{\alpha=4}^{6} E_\alpha \quad , \tag{7.3}$$

where x and y are defined by (6.4) and (6.5), respectively; w and E_4 to E_6 are given by

$$w = \sum_{i,\mu} a^*_{i\mu} D_{i_\mu} + \sum_{i,\mu} D^*_i a_{i_\mu\mu} - 2y \quad , \tag{7.4}$$

and

$$E_\alpha = \sum_{i,\mu\neq\mu'} a^*_{i,\mu'} a_{i_\mu\nu_\alpha} V_\alpha^{i,i^\mu} \quad , \qquad (\alpha=4,5,6) \quad . \tag{7.5}$$

Limits on the energy can be estimated by the variational principle [5]. The bounds of x and y are given by (6.6) and (6.7) respectively, while we have

$$-6 \leq w \leq 6 \quad , \tag{7.6}$$

$$-3V_4^c \leq E_4 \leq V_4^c \quad , \quad \text{for} \quad y > 0, \tag{7.7a}$$

$$-V_4^c \leq E_4 \leq 3V_4^c \quad , \quad \text{for} \quad y < 0 \quad , \tag{7.7b}$$

and for $\alpha=5$ and 6 we have

$$-3|V_5^c| \leq |E_5| \leq V_5^c \quad , \qquad \text{for} \quad y > 0, \tag{7.8a}$$

$$-V_5^c \leq |E_5| \leq 3|V_5^c| \quad , \qquad \text{for} \quad y < 0 \tag{7.8b}$$

where the superscript c stands for the crystalline ($\Phi=60°$, $180°$) configurations. Note that the valence band has positive y values (bonding), while the conduction band has negative y values (antibonding).

With a view to finding out the effects of dihedral angle disorder and the ring statistics independently of the effects of bond-length and bond-angle disorder, let us study the case where $\Delta r/r = \Delta\theta/\theta = 0$. The V_1 and V_2 are invariant σ interactions. Note that $(V_4 + V_5 + V_6)$ is also a σ interaction. Besides, when $\Delta r = \Delta\theta = 0$ is satisfied and $V(\Phi)$ takes the form (7.2), it is easy to show that $(V_4 + V_5 + V_6)$ is independent of the dihedral angle and therefore invariant.

[Bottom of the Valence Band]

The bottom of the valence band is given by

$$E_V^{min} = 4V_1 + V_2 + 6V_3 + 3(V_4 + V_5 + V_6) \quad . \tag{7.9}$$

As $y = 4$, Ψ is a pure s state so that only σ interactions enter (7.9). It is insensitive both to dihedral-angle and to topological disorder. This limit is a normal band edge, and little difference is expected from the crystalline case.

[*Top of the Valence Band*]

The top of the valence band occurs at

$$E_V^{max} = V_2 + 2V_3 + 3V_4^c - V_5^c - V_6^c \quad . \tag{7.10}$$

As $y = 0$, Ψ is a pure p state. The σ component of the p state is highly sensitive to topological disorder, as discussed below, and the π component of the p state is highly sensitive to dihedral-angle disorder, as indicated by the presence of V_4, V_5, and V_6 in a noninvariant form. The sensitivity of the component is so great, in fact, that the density of states at the top of the valence band is dominated by the π states to a greater degree even than in the crystal. Thus the energy E_V^{max}, (7.10), is a band limit like that occurring in the alloy problem discussed by Lifshitz. Such energies can be approached only by states localized to those regions of the material having a crystalline configuration. Within those regions the states are of the crystalline Γ_{25}, character. The width of the tail of localized states can be estimated from a simple variational expression for

the position of the mobility edge, $E_V^{max} - E_C$ (mobility edge) $\gtrsim \overline{E_V^{max}(\phi)}$ which requires a knowledge of the probability distribution of the dihedral angles. Structural models give approximately the distribution

$$P(\phi) \propto (\frac{2}{3} \sin^2 \frac{3}{2}\phi + \frac{1}{3}) \quad . \tag{7.12}$$

Eq.(7.10) and the variational expression (7.11) give an estimate of 0.3 eV to the width of the tail. Recent experimental data indicates the dihedral-angle distribution quite different from (7.12)[38], but whatever the distribution may be, the width of the tail is expected to be of the order of a few tenth of an eV, since it cannot be larger than 1.0 eV.

[*Bottom of the Conduction Band*]

The bottom of the conduction band is more difficult to analyze, and we have found only strong plausibility arguments that its energy is the energy at X_1 in the crystal and the bottom of the crystal conduction band in this model. We note that the energy is insensitive to dihedral-angle disorder. The wave function is of X_1 type and can be represented by pure s and pure pσ functions alternating across bonding directions. Thus, it cannot be achieved on an odd ring, as for the antibonding state in the s-band problem. There will instead be frustrated bonds on which sp-σ interactons are replaced by pp-σ interactions, a relatively unimportant effect. Thus the tail states of the conduction band are somewhat sensitive to topological disorder and localized to those regions of the material containing only even rings. The energy (7.9) marks a Lifshitz-like limit. Using a figure of 12% of the bonds being frustrated [cf. 39], a variational estimate for the position of the mobility edge yields 0.1 eV for the width of the tail.

Table 5 Effects of various kinds of disorder on the properties of the band edges. Notation ⊚ denotes 'strongly influenced'; ◯ 'reasonably influenced' and X 'rather insensitive'.

			DAD	RING DISORDER	BOUND
CONDUCTION BAND	BOTTOM (IF X_1)		X	⊚	LIFSHITZ LIMIT
VALENCE BAND	TOP	$P(\sigma)$	◯	⊚	LIFSHITZ LIMIT
		$P(\pi)$	X	X	
	II + III		X	◯	
	BOTTOM		X	X	NORMAL EDGE

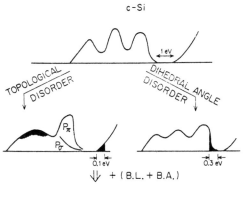

c-Si

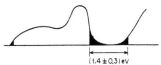

$\Downarrow$ + (B.L. + B.A.)

(1.4 ± 0.3) eV

Fig.14 Schematic illustra-tion of effects of various kinds of disorder on the behavior of the DOS. ('BLD' stands for the bond-length disorder while 'BAD' for the bond-angle disorder).

141

A summary of the investigations in this section is given in Table 5 and in Fig.14.

8. Discussion

To summarize, we have learned [5] that, in a tetrahedrally-bonded ideal CRN,

(1) the effects of the bond-length and bond-angle disorder are not very important, the energy shift due to the former being $\sim\pm0.02$ eV while the shift due to the latter is about ±0.2 eV;

(2) the valence-band top is sensitive to the dihedral-angle disorder and the estimated tail width is about 0.3 eV;

(3) the conduction-band bottom, if it occurs at X_1 point, is influenced by the ring disorder, the estimated tail width being about 0.1 eV.

Therefore, the upper bound of the mobility gap is 1.7 eV where the gap for a crystal is taken to be 1 eV.

It is clear from the discussion in the preceding sections that both the nature of the wavefunctions near the band edges and the magnitude of disorder are important in determining the widths in the band tails. Dihedral-angle disorder can be reduced by adding atoms of an element with a large electronegativity difference from silicon (or another host amorphous semiconductor), e.g. F added to Si [1]. This increases the energy barrier between the eclipsed and staggered positions, thus resulting in a narrow tail width at the top of the valence band. For example, the barrier for rotation around Si-F is ~70 MeV or ten times as large as that for Si-H. This would generally improve the short-range order in a-Si:FH over that on a Si:H or a-Si, as proposed by Ovshinsky [private communication], and explain the observed narrowing of the valence-band tail over that of a-Si:H [1].

The odd-even disorder can be suppressed by using III-V compounds. If coordination defects can be suppressed in these materials, they could give semiconductors with very good conduction band edges. As discussed in a preceding section, the ideal material would be one with the bottom of the conduction band near X_1.

It is certainly to be expected that the inclusion of F or the construction of III-V compounds introduces quantitative disorder such as bond-length disorder and bond-angle disorder. Our assertion however is; let us be optimistic and assume that these effects might be unimportant compared to the effects explained in the above two paragraphs and try to make the above-proposed materials to see what happens.

In this paper, we have studied the electronic properties of "ideal amorphous silicon or germanium" which we have defined as a structure with a fully interconnected continuous random network (CRN) of four-fold atoms. The only information which the above definition conveys about the structure is that the coordination number is four. Any other parameters are left arbitrary. Actually, in order to describe the structure completely, it is necessary to give all many-atom distribution functions $g^{(n)}(R_1,\cdots,R_n)$. But, let alone the many-atom distributions, the above definition of "ideal amorphous silicon" carries no information about bond-length, bond-angle, dihedral-angle distributions and the ring statistics. Therefore, a question arises: is there more than one set of distributions of these parameters or not? or in other words is there more than one ideal amorphous structure or is there only one unique ideal amorphous structure? Of course, in the more quantitative aspects of the work reviewed above, we took the bond length, bond angle, and dihedral angle distribution as those appropriate to real a-Si.

It is interesting to note in this connection that, although the original definition of 'glass' is an amorphous solid prepared from the melt through the glass transition (see §2), there has been a recent tendency to classify as glasses the kinds of amorphous solids which have not been prepared through the melt but which still show comparatively stable properties [40]. Then, again we ask; has the concept of glass in this context something to do with the concept of ideal amorphous structure?

As mentioned in §6, ideal, pure a-Si may exist in concept only. Therefore, it may be appropriate to extend the concept of ideal, pure a-Si to include those in which all coordination defects are 'killed' by the alloy atoms H or F. Then, "the extended ideal amorphous structure" can be connected to real materials, of which various observations are possible, and therefore we can hope to obtain some detailed information of structure from experiment. Among other things, it is important to know whether or not the parameters describing short-range and intermediate-range structures are dependent on preparation.

Although disordered systems in general have been examined theoretically to quite an extent, the theoretical study of amorphous semiconductors is still in its infancy and we are left with wide range of problems to be solved.

References

[1] M.H. Cohen, J. Singh and F. Yonezawa: J. Non-Cryst. Solids 35 & 36 55 (1980)
[2] M.H. Cohen, H. Fritzsche, J. Singh and F. Yonezawa: to be published in Proc. 15th Int. Conf. on the Physics of Semiconductors, Kyoto, Sept. 1-5 (1980)
[3] M.H. Cohen, J. Singh and F. Yonezawa: Solid State Communications (in press)
[4] J. Singh and M.H. Cohen: submitted to Phys. Rev. Letters
[5] J. Singh: submitted to Phys. Rev.
[6] M.H. Brodsky: *Topics in Applied Physics Vol.36 "Amorphous Semiconductors"* (Springer-Verlag, Berlin 1979)
[7] J.M. Ziman, *"Models of Disorder"* (Cambridge University Press, London 1979)
[8] A.F. Ioffe and A.R. Regel: Progr. Semiconductors 4 239 (1960)
[9] D. Adler: *"Amorphous Semiconductors"* (CRC Press, Ohio 1971)
[10] W.E. Spear and P.G. LeComber: J. Non-Cryst. Solids 8-10 727 (1972); A. Madan, P.G. LeComber and W.E. Spear: J. Non-Cryst. Solids 20 239 (1976)
[11] S.R. Ovshinsky and A. Madan: Nature 276 482 (1978)
[12] D.S. Saxon and R.A. Hutner: Philips Res. Reports 4 81 (1949)
[13] F. Cyrot-Lackmann: J. Phys. C5 300 (1972)
[14] C.E. Carroll: Phys. Rev. B12 4142 (1975)
[15] F. Ducastelle: J. Physique 35 983 (1974)
[16] I.M. Lifshitz: Advan. Phys. 13 483 (1964)
[17] R. Grigorovici: *Amorphous and Liquid Semiconductors* ed. J. Tauc (Plenum Press, London and New York, 1974) p.45
[18] R. Grigorovici and R. Manaila: Thin Solid Films 1 434 (1968)
[19] D.E. Polk: J. Non-Cryst. Solids 5 365 (1971)
[20] D.E. Polk and D.S. Boudreaux: Phys. Rev. Lett. 31 92 (1973)
[21] G.A.N. Connell and R.J. Temkin: Phys. Rev. B9 5323 (1974)

[22] N.J. Shevchik: Phys. Stat. Sol.(b) 52 K121 (1971)
[23] L. Ley, S. Kawalczyk, R. Pollak and D.A. Shirley: Phys. Rev. Lett. 29 1088 (1972)
[24] J.D. Joannopoulos and M.L. Cohen: Phys. Rev. B7 2644 (1973)
[25] M.F. Thorpe, D. Weaire and R. Alben: Phys. Rev. B7 3777 (1973)
[26] J.D. Joannopoulos: Phys. Rev. B16 5488 (1977)
[27] T. Hama and F. Yonezawa: Solid State Comm. 29 371 (1979)
[28] D. Weaire and M.F. Thorpe: Phys. Rev. B4 2508 (1971)
[29] M.F. Thorpe and D. Weaire: Phys. Rev. B4 3518 (1971)
[30] M.F. Thorpe and D. Weaire: Phys. Rev. Lett. 23 1581 (1971)
[31] D. Weaire: Phys. Rev. Lett. 26 1517 (1971)
[32] V. Heine: J. Phys. C: Solid State Physics 4 L221 (1971)
[33] J. Ziman: J. Phyc. C: Solid State Physics 4 3129 (1971)
[34] F. Yonezawa: *Proc. 5th Int. Conf. on Amorphous and Liquid Semiconductors*, ed. Stuke and Brenig (Taylor & Francis, London, 1974) p.829
[35] J.C. Slater and G.F. Koster, Phys. Rev. 94 1498 (1959)
[36] J. Chelikowsky and M.L. Cohen: Phys. Rev. B10 5095 (1979)
[37] F. Yonezawa: unpublished
[38] I. Solomon: private communication
[39] W.Y. Ching, C.C. Lun and L. Guttman: Phys. Rev. B16 5488 (1977)
[40] H. Fritzsche: this volume

Some Problems of the Electron Theory of Disorderd Semiconductors

V.L. Bonch-Bruevich

Faculty of Physics, Moscow University
Moscow, B-234, USSR

1. Introduction

In this talk two problems of the electron theory of disordered semiconductors
are considered. One of them is connected with the Anderson localization,
another one—with the nature of the random field acting upon the charge carriers in disordered semiconductors of various chemical composition.

The localization problem has often been a subject of discussion. The very
fact of the existence of discrete fluctuational levels seems nowadays to
raise no doubts. Yet, there are points concerning the localization criteria
which still seem to need some clarification. These are treated in Sects.
11.2,3 (following [1]).

The concept of a random force field proved to be rather helpful when inter-
preting a number of experimental data. However, the particular nature of the
field in various substances was not made quite clear until recently. This
problem is considered in Sects.11.4,5.

2. Localization Problem – The Rocks to be Avoided

When treating the fluctuational levels some complications are met which are
connected with the very nature of the physical system considered and which
have often been a source of misunderstandings and mistakes.

The first complication arises when one considers the type of wave functions describing the bound states (if any). In a conventional quantum-mech-
anical force center problem, discrete levels correspond to the wave functions which belong to the Hilbert space, L_2, and are localized near the
center. However, when turning to the problem in question this statement
needs to be clarified. The point is that we have to deal with a set of potential wells ("centers"), the well of some prescribed form and depths

("type") being repeated many times. In the thermodynamic limit both the number of particles, $\mathcal{N}$, and the volume of the system, Ω, tend to infinity while

$$0 < \lim_{\substack{\mathcal{N} \to \infty \\ \Omega \to \infty}} \frac{\mathcal{N}}{\Omega} = n < \infty \quad .$$

(1)

Under such conditions the number of potential wells of a given type tends as well to infinity, leaving the concentration of the relevant levels finite and nonzero. Thus the localization has to be understood in a somewhat conditional way: one has to consider a finite region of space containing just a single well of a certain type.[1] The region should be small compared to Ω, its volume staying finite at $\Omega \to \infty$. At the same time the region should be big enough (practically infinite) compared both with the atomic dimensions and with the localization radius; the latter condition makes sense provided the levels in question are not too shallow.

Thus the physical system considered should contain at least three scales of length. The first of them is a microscopic one, the relevant linear dimension being that of a typical localization radius. The second scale corresponds to the "localization regions" just described. The third one is macroscopic. It is defined by the dimensions of the sample; the self-averaging of the observable quantities takes place within the volumes of such a dimension. Note that the first and second scales are not absolute, but depend upon the energy of the electron.

To illustrate this argument it is convenient to write down, say, the one-electron Green's function, $G^{(+)}$ of the problem considered. Let Ψ_λ, W_λ and λ be the wave function, energy and a set of quantum numbers describing the electron stationary states in some configuration of a random field. Then the Fourier transform of $G^{(+)}$ with respect to time, defined as in [2], is given by

$$G^{(+)}(\underline{x},\underline{x}';E) = -\frac{1}{2\pi} \sum_\lambda \frac{\Psi_\lambda(\underline{x})\Psi_\lambda^*(\underline{x}')}{E - W_\lambda} \quad .$$

(2)

Here $\underline{x}$, $\underline{x}'$ and E are the spatial coordinates and energy argument respectively ($\hbar = 1$; spin coordinates are omitted for simplicity), E containing a conventional small imaginary part; sum over λ may in fact mean an integral if needed.

---

1 To avoid a misunderstanding it should be emphasized that the region may contain (moreover, does contain) quite a number of wells of other types, sustaining the levels differing in energy.

Strictly speaking, the very set of quantum numbers is random. However, with an overwhelming probability the system of interest possesses no symmetry except that with respect to the translations and rotations of the sample as a whole. Hence, in a majority of configurations the set λ, describing the bound states, contains just the energy $W_\lambda = W$, radius-vector of the localization center, $\underline{R}_\lambda = \underline{R}$ and, possibly, the spin quantum number $\sigma_\lambda = \sigma$; according to what had been said above there is a degeneracy with respect to $\underline{R}$. Thus the part of the Green's function corresponding to the discrete spectrum takes the form

$$G^{(+)}(\underline{x},\underline{x}';E) = -\frac{1}{2\pi} \sum_W \frac{1}{E - W} \sum_{\sigma,\underline{R}} \Psi_{W,\sigma}(\underline{x} - \underline{R}) \Psi^*_{W,\sigma}(\underline{x}' - \underline{R}) \quad . \tag{3}$$

The arguments of the wave functions have been written in the form $\underline{x} - \underline{R}$, $\underline{x}' - \underline{R}$ to emphasize that the functions are localized near the points $\underline{R}$.

It is seen that once the components of $\underline{x},\underline{x}'$ are confined within the region mentioned above, the sum over $\underline{R}$ contains in fact just one term of importance. In other words the situation is effectively the same as that met within the one-center problem.

The second complication is connected with the very definition of the discrete and continuous spectra. The point is that the concept of a continuous spectrum becomes operative only in the limit $\Omega \to \infty$; however, then the discrete levels may well form an everywhere dense set. To discriminate between the two parts of the spectrum one has to consider the wave functions belonging to separate potential wells: in the limit $\Omega \to \infty$ they either belong or do not belong to the class L_2. For convenience we keep the terms "continuous" and "discrete" at $\Omega < \infty$ as well keeping in mind the meaning just described.

Finally the third rock comes from a temptation to replace the sum over $\underline{R}$ in (3) by an integral over the sample. Then the Green's function would depend just upon the difference $\underline{x} - \underline{x}'$ and its diagonal elements, $G^{(+)}(\underline{x},\underline{x};E)$, would become position independent—with a "conclusion" of a spatially homogeneous distribution of electrons. However, it is obvious that such a replacement means taking an average over the coordinates of the localization centers. The variations of electron density taking place over a nonmacroscopic scale are then smeared out—in accordance with the macroscopic homogeneity of a sample. This however does not mean that such variations are indeed absent.

3. Localization Problem – Criteria

Various forms of the localization criteria were suggested many a time (see a review paper [3]). However, it did not seem always quite clear in what measure was account taken of the peculiarities of the problem mentioned in Sect.11.2. Therefore, it seems worthwhile to suggest yet another approach (which is equivalent to those known previously). It is convenient to consider the macroscopically homogeneous disordered semiconductor as a system degenerate in the Bogoliubov's sense [4]. To see whether there are localized states, one has to lift the degeneracy with respect to the coordinates of the localization centers. To this end one has to include a small localized perturbation in the Hamiltonian. We put

$$\mathcal{H} = \mathcal{H}_0 + \eta v(\underline{x} - \underline{x}_0) \quad . \tag{4}$$

Here $\mathcal{H}_0$ is the Hamiltonian of the physical system considered, including a random potential energy as well as the electron-electron interaction and the interaction of electrons with the vibrations of atomic matrix; η is an infinitely small quantity; $\underline{x}_0$ is some arbitrary chosen fixed point; and v is some function localized near the zero value of its argument. Explicit form of v is irrelevant; for convenience we put $x_0 = 0$ and

$$v = \delta(\underline{x}) \quad . \tag{5}$$

According to the rule formulated in [4], we have to calculate the variation of an electron density induced by the perturbation ηv. If the variation remains finite at $\eta \to 0$, the localized states are present, otherwise there are no such states.

Evidently it suffices to calculate the variation, δG, of the Fourier transform of some one-particle Green's function, $G(\underline{x}, \underline{x}'; E)$. The localization criterion may then be written in the form

$$\delta G(\underline{x}, \underline{x}; E) = O(\eta^0) \quad . \tag{6}$$

Let $\tilde{G}$ be a Green's function corresponding to the Hamiltonian (4):

$$\tilde{G} = G + \delta G \quad .$$

Equation for the Fourier transform of $\tilde{G}$ is

$$\tilde{G} = G + \eta G v \tilde{G} \quad . \tag{6'}$$

Using (5) we obtain

$$\delta G(\underline{x},\underline{x};E) = \eta \, \frac{G(\underline{x},0;E)G(0,\underline{x};E)}{1 - \eta G(0,0;E)} \, . \tag{7}$$

Thus, the criterion (6) is seen to be satisfied if the function $G(0,0;E)$ (still unaveraged over the random field) has the poles (not the branch cuts) on the real E-axis. This is the well-known result [5]. According to the first theorem on the correlation between the density of states and the electrical conductivity [6], such a criterion is equivalent to that suggested by Mott [7]: the states are localized if at $T \to 0$ the electrons contained therein give a zero contribution to the static conductivity. Conceptually, Mott criterion is more convenient since it deals with a directly observable quantity averaged over the random field. On the other hand this criterion leads to serious complications if numerical methods are used. Indeed, for obvious reasons, the static conductivity may be nonzero only in the thermodynamic limit (1). On the other hand a computer calculation has to do with a finite sample. Probably this was one of the reasons which stimulated the attempts to replace the Mott criterion by some others which would be equivalent to it in the limit $\Omega \to 0$, but would as well make sense at $\Omega < \infty$ [8-10]. Having this in mind it seems worthwhile to consider yet another possibility suggested by a behaviour of a real part of the conductivity in some energy region E at a finite frequency ω. Then the result for a finite system should be nonzero; the localization criterion would be reduced to studying the sign of the derivative of the quantity considered with respect to the frequency: $\sigma = \sigma_1 + i\sigma_2\sigma'_{1,\omega} > 0(< 0)$ if the states are localized (delocalized) in the sense of Sect.11.1.

4. Intrinsic Random Field in Disordered Semiconductors[2]

Intrinsic random field in amorphous and liquid semiconductors may be produced both by impurities or other imperfections and by an "intrinsic" disorder, the latter being present even in an ideal glass. In this section the fields of the latter origin are considered. A random field is conveniently described by a binary correlation function

$$\underline{\psi}(\underline{x},\underline{x}') = <\delta v(\underline{x})\delta v(\underline{x}')> \, . \tag{8}$$

---

2 This section is based upon [11,12].

Here the angular brackets mean averaging over the random field, $\delta V = V - \langle V \rangle$ is the electron potential energy fluctuation.[3] To calculate the r.h.s. of (8) it is convenient to replace the potential by a pseudopotential. Consider first a homopolar material assuming the pseudopotential to be local and energy independent. Then

$$V(\underline{x}) = -\sum_{i=1}^{\mathscr{N}} v(\underline{x} - \underline{R}_i) \quad , \tag{9}$$

where $v(\underline{x} - \underline{R}_i)$ is the pseudopotential produced by an atom placed at the point $\underline{R}_i$. The index i labels the atoms the average concentration of which is $n = \mathscr{N}/\Omega$.

Let $g(r)$ be the atomic radial distribution function. Its explicit form is irrelevant for what follows. Important is just the fact that, only the short range order being present, $g(r) - 1 \to 0$ at $r > r_c$; here r_c is of the order of an average interatomic distance. Let the Fourier transform of v be defined by

$$v(q) = n \int d\underline{x} \, e^{-i\underline{q}\underline{x}} \, v(\underline{x}) \quad .$$

Then one obtains in a straightforward way

$$\underline{\psi}(r) = \frac{n^{-1}}{2\pi^2 r} \int_0^\infty q \, S(q) v^2(q) \sin(qr \, dq) \quad . \tag{10}$$

Here

$$S(q) = 1 + n \int dr[g(\underline{r}) - 1] e^{-i\underline{q}\underline{r}} \quad . \tag{11}$$

This is a well-known interference function; its values may be directly obtained from experimental data on the scattering of X-rays, electrons, and neutrons.

Equations (10,11) are easily generalized to the case of a compound. Let a, a' label the types of the atoms (ions). Then we have to deal with a set of functions $g_{aa'}(\underline{r})$ and, accordingly, with a set

$$S_{aa'} = \delta_{aa'} + \sqrt{n_a n_{a'}} \int dr[g_{aa'}(\underline{r}) - 1] e^{-i\underline{q}\underline{r}} \quad . \tag{12}$$

The correlation function of a random field is now given by

---

3 In what follows we consider macroscopically homogeneous and isotropic systems. Then $\underline{\psi}(\underline{x},\underline{x}') = \underline{\psi}(|\underline{x} - \underline{x}'|)$ and $\langle \delta v \rangle$ is coordinate independent.

$$\underline{\psi}(\underline{r}) = \sum_{a,a'} \frac{(n_a n_{a'})^{-\frac{1}{2}}}{(2\pi)^3} \int d\underline{q}\ S_{aa'}(\underline{q}) v_a(\underline{q}) v_{a'}(-\underline{q}) e^{i\underline{q}\underline{r}}\ . \tag{13}$$

Since the atomic pseudopotentials are available for a number of substances (10) and (13) may be used to obtain $\underline{\psi}(r)$ numerically from experimental data. However, in the two cases of importance an analytical treatment is sufficient.

Consider first a material with short-range forces. Studying the asymptotics of the r.h.s. of (10) shows that at $r \gg r_c$ the correlation function drops rapidly at any reasonable form of a pseudopotential. Hence, if the lengths exceeding r_c are of interest we may put

$$\underline{\psi}(r) = \mathscr{V}_0 \delta(\underline{r})\ . \tag{14}$$

Such a form was often used on phenomenological grounds. Equation (10), however, allows $\mathscr{V}_0$ to be expressed in terms of directly measurable quantities:

$$\mathscr{V}_0 = n^{-1} v^2(0) S(0)\ . \tag{15}$$

As shown by Mironov [13], in the case (14) the characteristic energy $\bar{w}$ determining the scale of both the density of states and optical tails is given by[4]

$$\bar{w} = (16\ m^{3/2}\ \hbar^{-3}\ \mathscr{V}_0)^2\ . \tag{16}$$

Here m is an electron (or hole) effective mass corresponding to the auxiliary periodic field problem [14]. Consider an amorphous silicon using the data for $S(q)$ obtained in [15]. Then one obtains from (15) and (16): $\bar{w} \simeq 10^{-3} \div 10^{-2}$ eV. This seems to explain practical absence of the tails in thoroughly annealed samples of amorphous silicon. However, the situation may change in presence of impurities or other imperfections, including some "semimacroscopic" imperfections of a technological origin — see the next section. Thus measuring the light absorption coefficient in the tail region may be used to control the quality of the homopolar amorphous materials considered.

Now we turn to the materials where the bonds are — at least partly — of a polar nature. Such seem to be, in particular, some chalcogenide glasses. Being again interested in the cases when the lengths of importance exceed

---

4 An analogous result was obtained (by another method) by M.H. Cohen, J. Singh, and F. Yonezawa [16].

the interatomic distances, we may take $v_a(q)$ as just a screened potential
of an ion with an effective charge $Z_a e$. The screening may be due to any of
the mechanisms well-known in semiconductor physics. We use the screening
radius r_0 as a phenomenological quantity, subject just to an obvious condition: $r_0 \gg r_c$. In this way we obtain, neglecting the quantities of the
order of r_c/r_0:

$$\underline{\Psi}(r) = \frac{2\pi r_0 e^4}{\varepsilon^2} \exp\left(-\frac{r}{r_0}\right) \sum_a n_a Z_a^2 \quad . \tag{17}$$

Formally this is just the binary correlation function for the Coulomb random
field produced by a set of randomly spaced point charges. An effective concentration of such "ions" is

$$n_t^* = \sum_a n_a Z_a^2 \quad . \tag{18}$$

Since n_a^{-1} is an atomic volume, the r.h.s of (18) may be of the order of
10^{21} cm^{-3}. Thus, the theory of heavily doped semiconductors may be applied
to the materials considered the role of "charged impurities" being played
by irregularly spaced atoms of the substance itself. In particular the optical tail is obtained described by the Urbach rule. The characteristic energy
$\bar{w}$ is now given by [17][5]

$$\bar{w} = 1.1 \frac{m_n e^4}{\varepsilon^2 \hbar^2} \left[n_t^* \left(\frac{\varepsilon \hbar^2}{m_n e^2} \right)^3 \right]^{2/5} \quad . \tag{19}$$

This equation is valid at $m_n \ll m_p$. The r.h.s. of (19) may well be of the
order of 0.1 eV. Thus an intrinsic random field due to just the absence of
the long-range order may produce quite considerable tails in the materials
considered.

5. On the Nature of a Smooth Random Field

A random field is called smooth [17] if

$$\frac{\hbar^2 \Psi_2}{4m\Psi_1^{3/2}} \ll 1 \quad , \tag{20}$$

---

5 In contrast to the materials with short-range forces the energy $\bar{w}$ describes
  now just the optical tail, but not the density of states tail.

where $\Psi_1 = \langle(\delta V)^2\rangle$, $\Psi_2 = \frac{1}{2}\langle(\nabla\delta V)^2\rangle$. Such a field may arise, for example, in a system with Coulomb forces provided a set of point charges may be replaced by a continuous distribution of the space charge density, i.e., provided an approach of macroscopic electrodynamics is justified.[6] Such a situation is realized, in particular, in the samples containing some large scale random variations of impurity concentration or of other imperfections (not necessarily point ones). The imperfections may be of a technological origin or produced by irradiation. An example of such a random field was considered in [11].

Another reason for a smooth random field to arise is of a more fundamental nature: the concept of such a field may just reflect the well-known [18-20] fact of a smooth random bending of the bands taking place in disordered semiconductors.

Irrespective of its origin, a smooth field is conveniently characterized by the parameters Ψ_1 and Ψ_2. The latter may depend upon the band—conduction band or valence band— to be considered. This is the case when the field is produced by a deformation potential due to random strains in the material.

References

1. V.L. Bonch-Bruevich: Problems of the electron theory of disordered semiconductors. Preprint of the Institute of Theoretical Physics. Acad. Sci. of the Ukranian SSR - 78 - 67 R. Kiev (1978)
2. V.L. Bonch-Bruevich, S.V. Tyablikov: The Green Function Method in Statistical Mechanics. Translated by D. Ter Haar. North-Holland Publ. Comp. Amsterdam 1962
3. D.J. Thouless: Phys. Reports C13, N 3, 93 (1974)
4. N.N. Bogoliubov. Quasiaverages in the Problems of Statistical Mechanics (in Russian). In a book "Statisticheskaya Fizika i Kwantowaya Teoriya Polya", "Nauka", Moscow (1973)
5. E.N. Economon, M.H. Cohen: Phys. Rev. Lett. 25, 1455 (1970); Phys. Rev. B5, 2931 (1972)
6. V.L. Bonch-Bruevich, A.G. Mironov, I.P. Zviagin: Riv. del Nuovo Cim. 3, N 4, 321 (1973)
7. N.F. Mott, E.A. Davis: Electronic Processes in Non-Crystalline Materials. Clarendon Press Oxford (1971)
8. J.T. Edwards, D.J. Thouless: J. Phys. C5, 807 (1972)
9. D.W. Weaire, B. Kramer: Journ. Non-Cryst. Sol. 35-36, 9 (1980)
10. F. Yonezawa: Journ. Non-Cryst. Sol. 35-36, 29 (1980)
11. V.L. Bonch-Bruevich, V.D. Karaivanov, Ja.G. Proikova: Phys. Stat. Sol. (b) 96, 271 (1979)

---

6 Nonrandom fields of such a type are well-known. They arise near various junctions.

12. V.L. Bonch-Bruevich: Journ. Non-Cryst. Sol. *35-36*, 95 (1980)
13. A.G. Mironov: Vestnik Moskovskogo Universiteta, Fizika i Astronomiya (in press)
14. V.L. Bonch-Bruevich: Journ. Non-Cryst. Sol. *4*, 410 (1970)
15. N.C. Halder, R.L. Wourns: Z. Naturforsch. *30*, 55 (1975)
16. M.H. Cohen, J. Singh, F. Yonezawa: Journ. Non-Cryst. Sol. *35-36*, 55 (1980)
17. V.L. Bonch-Bruevich: Phys. Stat. Sol. *42*, 35 (1970)
18. H. Fritzsche: Journ. Non-Cryst. Sol. *6*, 49 (1971)
19. B.I. Shklovskii, A.L. Efros: GETP *62*, 1156 (1972)
20. V.L. Bonch-Bruevich: GETP *61*, 1168 (1971)

The Anderson Localisation Problem

D. Weaire

Physics Department, University College
Dublin, Ireland

Abstract

The current status of the Anderson localisation problem is reviewed. The various contributions of analytical and computational techniques are summarized.

1. Introduction

In the interpretation of transport data for amorphous semiconductors the concept of Anderson localisation has often played a key role. The valence and conduction band edges are considered to have the (schematic) form shown in Fig.1, with <u>mobility edges</u> separating extended and localised states, the latter making no contribution to (T=0) conductivity.

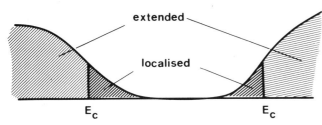

E_c E_c

<u>Fig.1</u> Sketch of the density of states around the gap of an amorphous semiconductor, with mobility edges at E_c, E_c'.

This picture is due largely to Mott [1]. The localisation of electrons by disorder has been studied by Anderson [2]. His conclusion that such localisation would indeed occur is not at all obvious or trivial - indeed it was at one time controversial. We may picture an exceptional fluctuation in the Hamiltonian binding a localised state in its vicinity, but will not other such states, however distant they may be, hybridise with it to form an extended state? If we think of the case of the centre of the band of states produced by a simple Hamiltonian, it is even less clear that disorder can localise the eigenstates, yet this is what Anderson claimed. Above a certain critical strength of disorder, <u>all</u> states are localised. It is towards this proposition that most theoretical studies have been directed. Their first goal has been to locate the critical strength of disorder for this <u>Anderson transition</u>, which provides the most convenient point of comparison of the results of different methods.

For simplicity the following Hamiltonian has generally been used.

$$H = \sum_j \varepsilon_j |j><j| + V \sum_{j,j\Delta} |j><j\Delta| \qquad (1)$$

The basis states j are located on the sites of a periodic structure, and are coupled by constant nearest neighbour interactions (V). Disorder is introduced in the form of fluctuating diagonal elements, or site energies (ε_j). These are random variables with a distribution which is usually chosen to be uniform between two limits, that is, the rectangular distribution,

$$p(\varepsilon) = 1/W, \qquad |\varepsilon| < W/2$$
$$= 0, \qquad |\varepsilon| > W/2 \qquad (2)$$

The strength of disorder is then represented by the width W, or rather W/V.

In this case we see the problem as a competition between the two terms of (1), which would in isolation produce localised (single site) and extended (Bloch) states respectively. Which wins?

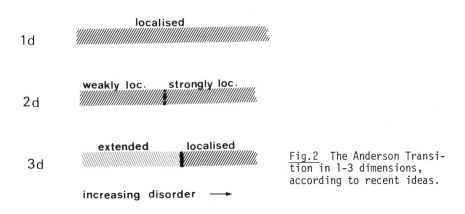

1d localised

2d weakly loc.   strongly loc.

3d extended   localised

increasing disorder ⟶

Fig.2 The Anderson Transition in 1-3 dimensions, according to recent ideas.

Fig.2 summarises the present consensus (which is not unanimous!) regarding the answer to this question in one, two and three dimensions. We see that Anderson's assertion has survived in three dimensions, but not in two. (The latter case is realised in, for example, inversion layers [3]). This has come as a surprise since the distinction between the two cases was not apparent in early work. The case of one dimension has always been regarded as a special one, with a number of fairly rigorous results supporting the statement that all states are localised for arbitrarily small disorder [4]. On the other hand, our understanding of two and three dimensions remains somewhat tentative. There have been practically no rigorous proofs of <u>any</u> non-trivial results for these cases. Recent scaling arguments may in due course be seen to be trustworthy and satisfy at least a physicist's standard of rigour. For the present they rest on a number of plausible but unproven assumptions and can hardly be considered invulnerable.

Without a star to steer by (such as was provided by the exactly soluble 2d Ising model in phase transition theory), theorists have wandered far and wide in their speculations, and reached a variety of conclusions. To test these, numerical calculations have been undertaken. Although they have generally proved "noisy" and difficult to interpret, these calculations have illuminated at least some features of the problem. They have shown, for example, that Anderson's original quantitative estimates of the critical disorder were very inaccurate.

In the remainng sections, we shall sketch the history of the analytical and numerical work which has led us to the picture given in Fig.2 for two and three dimensions, and comment further on its relevance to amorphous semiconductors.

2. Series methods

Anderson's original analysis [2], which is a subtle one in many respects, is based on the idea that the localisation of states should be reflected in the analytic properties of the Green's function. For a <u>periodic</u> Hamiltonian this is real on the real (energy E) axis, provided E is outside the allowed bands, while there is a branch cut, with non-zero imaginary part of G above/below the axis, whenever E lies in an allowed band. According to Anderson, the latter remains true in the disordered case <u>only for the range of extended states</u>. In the range of localised states the limit of $G(E+i\eta)$ for small η has zero imaginary part, except on a set of measure zero.

Anderson pursued this criterion via the study of series expansions for G, arguing that if a perturbation expansion (in V) converged, G must be real and states localised. The obvious perturbation expansion always diverges and it is necessary to rearrange it in a way which gives it a chance of convergence. Thouless [5] has clarified this and various other aspects of Anderson's treatment, at the same time emphasising its conjectural nature.

A formidable body of work [6] has been concerned with the gradual refinement of Anderson's approach, mostly in the sense of improving the approximations which one is forced to make in applying it to quantitative estimates of critical disorder. The questions regarding the correlation of terms in the perturbation series which lie at the heart of this work have proved difficult to resolve and should perhaps have been subjected to more scrutiny by numerical methods. While they may appear to have been set aside, they are likely to reappear in any detailed analysis of the currently fashionable scaling methods (section 5).

3. Numerical methods

Why not take the bull by the horns and simply calculate eigenvectors of (1) for some large array of sites?

This may be unkindly termed the brute force approach. In the outstanding example of a calculation of this kind, Yoshino and Okazaki [7] used a method of matrix diagonalisation which is very efficient for this application (at least in two dimensions) and obtained eigenvectors for systems of $10^2 \times 10^2$ sites. Thouless and co-workers [8,9] used a potentially more powerful approach, in which only eigenvalues were calculated, the degree of localisation being related to the sensitivity of the eigenvalues to changes in the boundary conditions.

Such calculations made it very clear that Anderson localisation was no illusion, since exponential localisation was clearly demonstrated for large disorder. It was even possible to locate the critical disorder with some confidence, at least in two dimensions. Yoshino and Okazaki went still further, in that they tried to evaluate the critical behaviour associated with the transition.

There have been various attempts to develop still more refined methods. The motivation for this lies in the observation that we are usually only looking for some average quantity, such as the average localisation length around a certain energy. It must be more efficient to calculate such averages in some direct way without generating a lot of redundant information. (At the same time it should be frankly admitted that Yoshino and Okazaki's calculations have hardly been surpassed by these more sophisticated techniques in the two-dimensional case!)

Weaire and Williams [10] showed how the equation-of-motion method, in which quantities of interest are extracted from the time dependence of a random wave vector, could be applied to the localisation problem. The quantity which emerges most naturally from this approach is the "inverse participation ratio", which is, roughly speaking, the inverse of the number of sites over which a localised eigenvector has significant amplitude. The application of the method was pursued by Weaire and Srivastava [11].

The recursion method has much in common with the equation-of-motion method and has also been applied to localisation [12].

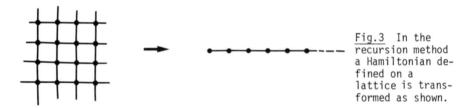

Fig.3 In the recursion method a Hamiltonian defined on a lattice is transformed as shown.

However it raises more difficult questions of interpretation, at least in our opinion [13]. The essence of the method is the transformation of the Hamiltonian into one which has the topological structure of a semi-infinite chain, and is hence amenable to a variety of simple techniques.

Yet another approach, which may be called simulation, was developed by Prelovsek [14]. It is similar to the equation-of-motion method, in that the time-dependent Schrödinger equation is to be integrated. However, Prelovsek simply takes a wave-packet which has a Gaussian envelope in space and consists of eigenvectors from a particular energy range, and watches it expand with time. If the states are extended it expands in the manner characteristic of diffusion $(r \sim t^{1/2})$. If the states are localised the wave packet remains localised.

All of these methods can be adapted to the calculation of the conductivity, in one way or another. In particular, the method of Thouless et al [8, 9] yields the conductivity rather directly (given certain assumptions), as does the simulation method, while the equation-of-motion and recursion methods may be used to evaluate the Kubo-Greenwood formula [15].

4. Numerical results

For diagonal disorder of the type specified above, all recent numerical calculations have found an apparent Anderson transition at

$$\frac{W_{crit}}{ZV} \simeq 1.6 \tag{3}$$

in the case of two dimensions. Here Z is the number of nearest neighbours, which provides a scaling factor such that different lattices can be approximately represented by the single critical value given in (3). Licciardello and Thouless [8] gave an argument to the effect that the minimum metallic conductivity (meaning the conductivity at the mobility edge) was a universal constant in two dimensions and their first numerical results seemed consistent with this. Later [9] more extensive calculations cast doubt on this conclusion. Moreover, as we shall see, recent ideas have caused a drastic revision of our picture of Anderson localisation in two dimensions.

It is much more difficult to perform satisfactory calculations in three dimensions. Most results point to

$$\frac{W_{crit}}{ZV} \simeq 2.0 - 2.5 \tag{4}$$

and (much more tentatively) to a zero minimum metallic conductivity.

There have also been some calculations [16, 17] for off-diagonal disorder. These have shown that it is rather ineffective in causing localisation of states, except at the band edges, as one might expect and as had been predicted by approximate analytic theory [18].

5. Scaling theories

The analogy between the localisation transition and a phase transition has always been an appealing one. It is therefore not surprising that there have been repeated attempts to bring over to this subject the results or methods of phase transition theory. Claims that there is a direct correspondence to some particular type of phase transition (allowing a direct transcription of critical exponents) have remained controversial. Less controversial are the "scaling theories" which borrow only the general style of thinking associated with the renormalisation group approach to the analysis of phase transitions. Using this approach, Abrahams et al, [19], now known almost universally as the "Gang of Four", reached the startling conclusion that all states are localised in two dimensions for arbitrarily weak disorder. This particular cultural revolution may not be nearly as drastic as it first appears, since it is expected that states are only weakly localised until a particular disorder strength is reached. Presumably this corresponds (for E=0) to the value of W_{crit} found in numerical calculations. Of course, this means that there can be no "minimum metallic conductivity" in any straightforward sense, and it is a further consequence of the theory that even in three dimensions (where the Anderson transition

survives) there is no minimum metallic conductivity. In that case the conductivity is found to disappear as $\sigma \sim (E-E_c)^\gamma$ where $\gamma > 0$. Thus one of the
central concepts of the theory of amorphous semiconductors is in danger of
being invalidated!

The "Gang of Four" argument is in some respects a surprisingly simple one
but the following precis may still be helpful.

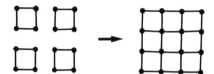

Fig.4 In scaling theory the coupling
of small blocks to form larger ones,
starting as shown, is considered.

Suppose we know the eigenvectors and eigenvalues of finite blocks (size L^d,
d=dimension) of our random system. If we couple these together, (as in
Fig.4) forming larger blocks with a larger length scale L, the eigenvectors
are coupled together by some matrix elements, of typical strength $\bar{V}$, say.
We might expect the ratio of this quantity to the spacing of the eigenvalues to be the appropriate measure of "strength of coupling". This is
the "g" parameter of [19]. Now the process which is envisaged will generate
larger blocks, which themselves have a g value, and it is a fundamental
assumption of this theory that this quantity depends only on the original g.
This means

$$\frac{d(\ln g)}{d(\ln L)} = \beta(g(L)) \tag{5}$$

and thus, for a given dimension, the function (g) is all that we need
know to determine how the coupling varies as L , and hence whether states
are localised. But how are we to know this function? An hypothesis due to
Thouless relates g to the conductivity , according to

$$g = \frac{2h}{e^2} \sigma L^{d-2} \tag{6}$$

and this enables the asymptotic behaviour of β to be established in the
limits $g \to 0, \infty$. Given this, it is argued that the only reasonable form for
the trajectories $\beta(g)$ are as shown in Fig.5, for d=1,2,3. A system which
starts with a given g must follow the arrows indicated. We see that in
two dimensions it can only scale towards weak coupling (localised states).

A remarkable feature of the theory is the ease with which it yields an interesting relation between the critical behaviour of the conductivity, on
one side of the Anderson transition, and the localisation length on the
other. Such a relationship is not at all obvious from other points of view,
although Mott once gave a rather intuitive argument for such a relationship
[20], with results not consistent with that now suggested by the scaling
theory.

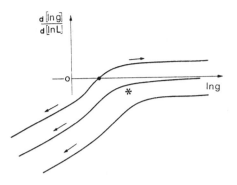

Fig.5 Scaling trajectories for 3d (top), 2d (middle) and 1d (bottom), according to Abrahams et al [19]. In 3d there is a fixed point ● and hence an Anderson transition for non-zero disorder strength, while in 2d there is only a smooth cross-over ✱ from weak to strong localisation.

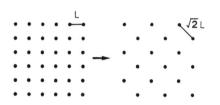

Fig.6 Decimation of the square lattice.

Scaling theories have a direct counterpart in numerical calculations, of which two categories may be distinguished. These are the "block coupling" and "decimation" techniques, each of which has already played a part in the study of phase transitions.

The block coupling approach was applied by Lee [21] to the two-dimensional case, with a view to checking the picture which had been presented by Abrahams et al. He examined the coupling of four finite blocks, as in Fig.4, by direct diagonalisation of the Hamiltonian matrix. By storing the eigenvalues and the boundary values of eigenvectors, he was then able to further couple the resulting larger blocks in a similar way, and so on, just as was envisaged in the original scaling argument. Such a procedure, if pursued with an interest in states at, say, E=0, may be made more efficient by discarding or approximating the small effects of eigenvectors with eigen-values far from zero, at every stage. The question then is, for a given strength of disorder — does the coupling of blocks go to zero (weak cou-pling limit) as the size of the blocks tends to infinity? This would indicate localisation. Lee's conclusion was that in two dimensions the variation of the scaling behaviour as the strength of disorder increased was more characteristic of the traditional (sharp) Anderson transition than the proposed smooth transition to weakly localised states. However, it may well be that this calculation suffers the same weakness as other numerical methods with regard to this question, namely, that it would need to be taken very far before the weakly localised states could be seen. For the moment it is to be regarded as preliminary and it will be most interesting to see what a more extensive calculation and analysis produces.

Decimation proceeds not by building up a large system but by taking it apart, as in Fig.6. One eliminates from consideration a complete sub-lattice of sites replacing them by effective interactions between the re-maining sites. These may then themselves be "decimated", and so on. The effective interactions should vary with increasing disorder in such a way as to scale towards the weak or strong coupling limits, indicative of localised and extended states.

Note that the off-diagonal interactions extend to all sites of the decimated system. They are found to decrease exponentially with distance at each stage of renormalisation, in the localised regime, as shown in Fig.7. Again, this provides opportunities for discarding the weakest parts of the interactions at each stage, but this has not been done in the preliminary studies so far undertaken [22, 23]. The potential of the method has therefore not been fully realised.

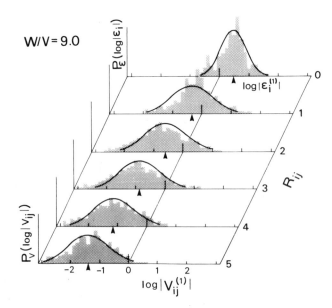

Fig.7 Probability distributions P_ε and P_V of renormalised site energies and off-diagonal interactions for vars distances R, as given by Aoki [22] for the square lattice after two renormalisations.

Such methods, which are still the object of study since their ideal modus operandi and precise interpretation is far from clear, offer the hope of the development of numerical techniques more powerful than those which have gone before, because of the feature of "discarding irrelevancies" to which we have referred.

6. Conclusion

The picture of Anderson localisation which is emerging contradicts the assumption which has often been made in analysing transport data for amorphous semiconductors, that the minimum metallic conductivity is finite. A notable example is the recent work of Spear et al [24], who have shown that a wide range of conductivity data (and various other measurements) for hydrogenated a-Si may be accommodated by such a model. However they require to postulate a mobility edge which moves rapidly with temperature. This is a rather puzzling concept, difficult to explain in terms of the temperature dependence of the strength of disorder, as these authors suggested. It would be more natural to attribute it to the failure of the low-T theory at higher temperatures, due to hopping in the localised states below the mobility edge - this point of view has been advocated by Döhler [25]. The theory may therefore need to be recast in two respects - the introduction of a zero minimum metallic conductivity, and the incorporation of effects due to hopping.

As for the estimation of the position of the mobility edges in a-Si, only Davies [26] has so far attempted this. While he obtains a result consistent with the observation, that is, a few tenths of an eV from the "band edge", the approximations and assumptions involved are very severe. (See also Yonezawa [27] for an analysis of this problem.)

Our conclusion must be that the Anderson localisation problem, posed in 1958 and a subject of growing interest since then, is still not quite solved in September 1980, and much remains to be done in the exploration of its relevance to amorphous semiconductors.

Acknowledgements

I wish to thank the staff of RIFP, Kyoto University for their hospitality and the Japan Society for the Promotion of Science for the award of a Visiting Fellowship.

References

1. N.F. Mott and E.A. Davis, Electronic Processes in Non-Crystalline Solids (Oxford UP, London, 1971)
2. P.W. Anderson, Phys. Rev. 109, 1492 (1958)
3. M. Pepper, Contemp. Phys. 18, 423 (1977)
4. K. Ishii, Suppl. Prog. Theor. Phys. 53, 77 (1973)
5. D.J. Thouless, J. Phys. C 3, 1559 (1970)
6. D.C. Licciardello and E.N. Economou, Phys. Rev. B11, 3697 (1975)
7. S. Yoshino and M. Okazaki, J. Phys. Soc. Japan 43, 415 (1977)
8. D.C. Licciardello and D.J. Thouless, J. Phys. C 8, 4157 (1975)
9. D.C. Licciardello and D.J. Thouless, J. Phys. C 11, 935 (1977)
10. D. Weaire and A.R. Williams, J. Phys. C 10, 1239 (1977)
11. D. Weaire and V. Srivastava, in Amorphous and Liquid Semiconductors, W. Spear ed., p.286 (CICL, University of Edinburgh (1977))
12. J. Stein and U. Krey, Z. Phys. B 37, 13 (1980)
13. D. Weaire and C.H. Hodges, J. Phys. C 12, L375 (1979)
14. P. Prelovsek, Phys. Rev. B 18, 3657 (1978)
15. D. Weaire and B. Kramer, J. Non-Cryst. Solids 32, 131 (1979)
16. D. Weaire and V. Srivastava, Solid State Commun. 23, 863 (1977)
17. K. Tsujino, A. Tokunaga, M. Yamamoto, and F. Yonezawa, Solid State Commun. 30, 531 (1979)
18. E. Economou and P.D. Antoniou, Solid State Commun. 21, 285 (1977)
19. E. Abrahams, P.W. Anderson, D.C. Licciardello and T.V. Ramakrishnan, Phys. Rev. Letters 42, 673 (1979)
20. N.F. Mott, Commun. Phys. 18, 423 (1977)
21. P.A. Lee, Phys. Rev. Letters 42, 1492 (1979)
22. H. Aoki, Solid State Commun. 31, 999 (1979); J. Phys. C 13, 3369 (1980)
23. D. Weaire and C.J. Lambert, to be published (1980)
24. W. Spear, D. Allan, P. Le Comber and A. Ghaith, J. Non-Cryst Solids 35/36, 357 (1980). See also W. Spear, these proceedings.
25. G.H. Döhler, J. Non-Cryst Solids 35/36, 363 (1980)
26. J.H. Davies, J. Non-Cryst Solids 35/36, 67 (1980)
27. F. Yonezawa, these proceedings.

Summary Talk

Hajimu Kawamura

Faculty of Science, Kwansei Gakuin University
Nishinomiya, 662, Japan

First of all, I would like to express my deepest appreciation to the
Organizing Committee of the Kyoto Summer Institute, especially to Prof.
Yonezawa who has arranged such a enjoyable Conference. I also thank to the
speakers who gave us so clear and stimulating lectures that even the begin-
ners in this field like me were much educated and excited.

I mentioned in the opening address of the International Conference on
the Physics of Semiconductors, that the amorphous semiconductors have been
attracting many scientists, because of the industrial requirement as well
as of the excitement to a spiritual adventure into a new frontier of science.
Indeed, I have found that the scientists meeting here are really adventurers
who are full of frontier spirit.

The introductory talk by Prof. Fritzsche was quite suggestive, fascinating and adventurous, though not sophisticated. The lecture by Prof. Solomon
reminded me of the closing address given by Prof.Aigrain at the Semiconductor
Conference in Paris in 1964. There, he mentioned that in good old day of
semiconductor physics, one can find something new, only with a small piece
of single crystal and simple experimental set up. This is still true in the
field of amorphous semiconductors. The talk by Dundee group again proved
this statement. I was amazed by the fact that they have figured out the
density of states in the energy gap with a quite simple set up, without using
such a sophisticated equipments as ultra high magnetic field, picosecond
laser, infrared and ultraviolet lasers and so forth.

It is rather surprising that there are still much controversies between
different reseach groups on the electrical and optical properties of a-Si,
perhaps because of the difference of the method of preparation, though the
solar cell made of hydrogenated a-Si is going to be used in a commercial
pocket computor. The problem of the surface and interfacial space charge
layers raised by Prof. Solomon is also controversial. The physics of the
hydrogenated a-Si is quite a young science, but the industry can not await
her ripening. This situation is somewhat similar to the situation of semiconductor physics in the 1940's, before and during the 2nd World War.
However, there is big difference. In the 1940's, we had a strong weapon.
That is Bloch theory. Unfortunately, we have no such a strong guiding principle in the physics of amorphous materials. The concept of Anderson localization may be a most powerful tool for the disordered system. But still the
amorphous is too awful object from the conceptual point of view. It was
impressive to me that the details of the density of states of the tetrahedrally bonded amorphous can be discussed from the tight-binding calculation
with great success. However, the situation is still far from predicting
the electronic behaviours in the actual amorphous semiconductors.

The study of the vibrational states seems to be little more advanced than that of the electronic states even in the hydrogenated and fluorinated silicon as shown by Prof. Lucovsky in his excellent talk.

In the chalcogenide glass which has lower coordination number with lone pair, the situation is much better compared with the a-Si which is tetrahedrally bonded and, therefore, highly constrained. Dr. Tanaka gave an interesting talk on the photoinduced structural change of chalcogen glass. In the chalcogenide glass, the melt-quenched glass and the well annealed evaporated film show the same macroscopic behaviour. This fact indicates that they have a single metastable state as mentioned by Prof. Fritzsche in his introductory talk. The reversible photo-structural change will be associated with the intermediate range order. In this connection I would like to refer to the talk given by Prof. Lannin in the seminar meeting. He showed that the static and dynamic correlations increase with the annealing in a-P. I think that this kind of work is important to clarify the structure of the glass in thermodynamically metastable state.

I shall conclude my personal and much biased summary with the following table. This is only my personal score showing the achievement in the

Achievements (%)

	material & structure	bonds	bands
crystal	99	80	90 (1980)
	30	5	10 (1940)
glass	30	20	10
amorphous	5	5	10

physics of semiconductors. From this table, you may predict that it will take another 40 years to achieve the same level of establishment as that of crystal semiconductors at present. I remember that Prof. Bonch-Bruevich told me at the last Kyoto Conference on Semiconductors in 1966, that the physics of disordered system will continue more than several ten years. However, if you take account of the fact that this field has been attracting so many active reseachers, explosive developments are expected in the near future.

Seminars Given During the KSI '80

The following seminars were given during the KSI '80. The details of these seminars are not given here.

Butcher, P.N.   Calculation of the Hall mobility of hopping carriers

Hamakawa, Y.   Why are amorphous semiconductors so interesting?

Hirabayashi, I.   Time-resolved luminescence and its fatigue effect in hydrogenated a-Si

Hirose, M.   Internal photoemission in hydrogenated a-Si

Imura, T.   Compositional studies of amorphous silicon films by Rutherford Backscattering spectrometry

Lang, D.V.   Advances in deep level transient spectroscopy: Application to hydrogenated a-Si

Lannin, J.S.   Static and dynamic correlation effects in amorphous semiconductors

Nemanich, R.J.   Compositional anisotropy and microstructure in a-Si:H

Nitta, S.   Optical properties of GDa-Si

Shimizu, T.   Effects of transition metal additives in chalcogenide glasses

Silver, M.   Recent simulation results on dispersive transport

Tsuji, K.   Structural studies of plasma-deposited amorphous Si:H

Tsujino, K.   Are the states really localized in two-dimensional disordered systems?

Weinstein, B.A.   Effect of pressure on the luminescence in hydrogenated a-Si

Photograph of the Participants of the Seminar

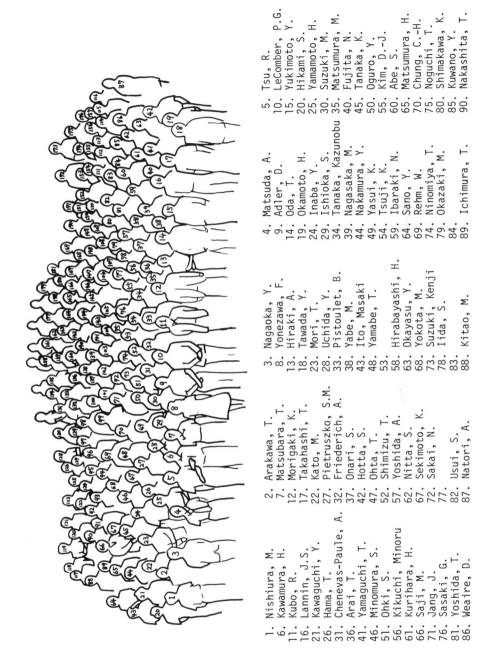

1. Nishiura, M.
2. Arakawa, T.
3. Nagaoka, Y.
4. Matsuda, A.
5. Tsu, R.
6. Kawamura, H.
7. Matsubara, T.
8. Yonezawa, F.
9. Adler, D.
10. LeComber, P.G.
11. Kubo, R.
12. Morigaki, K.
13. Hiraki, A.
14. Oda, T.
15. Yukimoto, Y.
16. Lannin, J.S.
17. Takahashi, T.
18. Tawada, Y.
19. Okamoto, H.
20. Hikami, S.
21. Kawaguchi, Y.
22. Kato, M.
23. Mori, T.
24. Inaba, Y.
25. Yamamoto, H.
26. Hama, T.
27. Pietruszko, S.M.
28. Uchida, Y.
29. Ishioka, S.
30. Suzuki, M.
31. Chenevas-Paule, A.
32. Friederich, A.
33. Pistoulet, B.
34. Tanaka, Kazunobu
35. Matsumura, M.
36. Arai, T.
37. Onari, S.
38. Yabe, M.
39. Nagasaka, M.
40. Fujita, N.
41. Yamaguchi, T.
42. Hotta, S.
43. Ito, Masaki
44. Nakamura, Y.
45. Tanaka, K.
46. Minomura, S.
47. Ohta, T.
48. Yamabe, T.
49. Yasui, K.
50. Oguro, Y.
51. Ohki, S.
52. Shimizu, A.
53.
54. Tsuji, K.
55. Kim, D.-J.
56. Kikuchi, Minoru
57. Yoshida, A.
58. Hirabayashi, H.
59. Ibaraki, N.
60. Abe, S.
61. Kurihara, H.
62. Nitta, S.
63. Okayasu, Y.
64. Sano, Y.
65. Matsumura, H.
66. Saji, M.
67. Sekimoto, K.
68. Yokota, M.
69. Rehm, W.
70. Chung, C.-H.
71. Jang, J.
72. Sakai, N.
73. Suzuki, Kenji
74. Ninomiya, T.
75. Noguchi, T.
76. Sasaki, G.
77.
78. Iida, S.
79. Okazaki, M.
80. Shimakawa, K.
81. Yoshida, T.
82. Usui, S.
83.
84.
85. Kuwano, Y.
86. Weaire, D.
87. Natori, A.
88. Kitao, M.
89. Ichimura, T.
90. Nakashita, T.

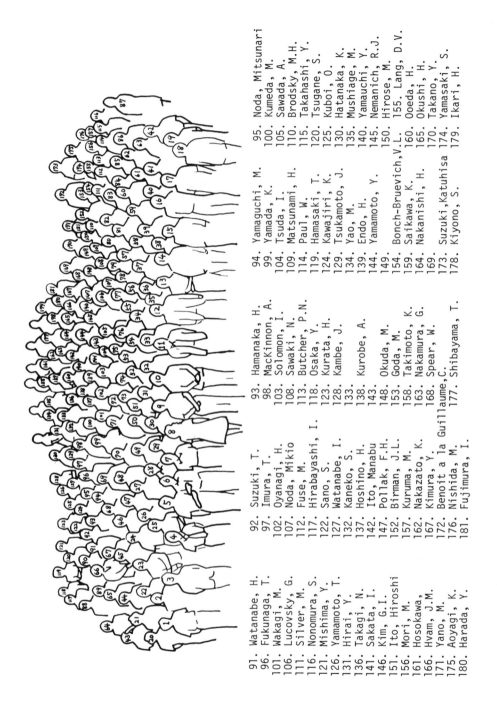

91. Watanabe, H.	92. Suzuki, T.	93. Hamanaka, H.
96. Fukunaga, T.	97. Imura, T.	98. MacKinnon, A.
101. Wakagi, M.	102. Oyanagi, H.	103. Solomon, I.
106. Lucovsky, G.	107. Noda, Mikio	108. Sawaki, N.
111. Silver, M.	112. Fuse, M.	113. Butcher, P.N.
116. Nonomura, S.	117. Hirabayashi, I.	118. Osaka, Y.
121. Mishima, Y.	122. Sano, S.	123. Kurata, H.
126. Yamamoto, T.	127. Watanabe, I.	128. Kambe, J.
131. Hirai, Y.	132. Kaneko, S.	133.
136. Takagi, N.	137. Hoshino, H.	138. Kurobe, A.
141. Sakata, I.	142. Ito, Manabu	143.
146. Kim, G.I.	147. Pollak, F.H.	148. Okuda, M.
151. Ito, Hiroshi	152. Birman, J.L.	153. Goda, M.
156. Mori, M.	157. Kuruma, M.	158. Takimoto, K.
161. Hosokawa,	162. Nakazato, K.	163. Nakamura, G.
166. Hvam, J.M.	167. Kimura, Y.	168. Spear, W.
171. Yano, M.	172. Benoit a la Guillaume,C.	
175. Aoyagi, K.	176. Nishida, M.	177. Shibayama, T.
180. Harada, Y.	181. Fujimura, I.	

94. Yamaguchi, M.	95. Noda, Mitsunari
99. Yamada, K.	100. Kumeda, M.
104. Tsuda, I.	105. Sawada, A.
109. Matsunami, H.	110. Brodsky, M.H.
114. Paul, W.	115. Takahashi, Y.
119. Hamasaki, T.	120. Tsugane, S.
124. Kawajiri, K.	125. Kuboi, O.
129. Tsukamoto, J.	130. Hatanaka, K.
134. Yao, M.	135. Mushiage, M.
139. Endo, H.	140. Yamauchi, Y.
144. Yamamoto, Y.	145. Nemanich, R.J.
149.	150. Hirose, M.
154. Bonch-Bruevich,V.L.	155. Lang, D.V.
159. Saikawa, K.	160. Ooeda, H.
164. Nakanishi, H.	165. Okushi, H.
169.	170. Takano, Y.
173. Suzuki,Katuhisa	174. Yamasaki, S.
178. Kiyono, S.	179. Ikari, H.

List of Participants

Number in *italics* refer to the person in the photograph on page 168

Abe, S. *(60)*   Inst. for Solid State Physics, Univ. of Tokyo,
                          Roppongi, Minato-ku, Tokyo 106, Japan

Adler, D. *(9)*   Room 13-3050, Massachusetts Inst. of Technology,
                          77 Massachusetts Ave, Cambridge, MA 02139, U.S.A.

Aiga, M.   LSI Research & Development Lab., Mitsubishi
                          Electric Corp., 4-1 Mizuhara, Itami, Hyogo 664, Japan

Aoyagi, K. *(175)*   The Inst. of Physical & Chemical Research
                          2-1 Hirosawa, Wako, Saitama 351, Japan

Arai, T. *(36)*   Inst. of Applied Physics, Univ. of Tsukuba,
                          Ibaraki 305, Japan

Arakawa, T. *(2)*   Technical Research Lab., Asahi Chemical Ind., Co.,
                          Ltd., 2-1 Samejima, Fuji, Shizuoka 416, Japan

Benoit a la Guillaume, C. *(172)* Groupe de Physique des Solides de L'E.N.S.,
                          Tour 23-2, Place Jussieu, 75221 Paris Cedex 05, France

Birman, J.L. *(152)*   Physics Dept., City College of City Univ. of New York,
                          New York, NY 10031, U.S.A.

Bonch-Bruevich, V.L. *(154)* Faculty of Physics, Moscow Univ.,
                          Moscow 117234, U.S.S.R.

Brodsky, M.H. *(110)*   IBM T.J. Watson Research Center, P.O. Box 218,
                          Yorktown Heights, NY 10598, U.S.A.

Butcher, P.N. *(113)*   Dept. of Physics, Univ. of Warwick, Coventry,
                          CV4 7AL, U.K.

Chenevas-Paule, A. *(31)* CEN de Grenoble, 85X 38041, Grenoble Cedex, France

Chung, C-H. *(70)*   Physics Dept., Yonsei Univ., Sodaemoon-ku,
                          Seoul 120-00, Korea

Davis, E.A.   Dept. of Physics, Univ. of Leicester, Leicester,
                          LE1 7RH, U.K.

Endo, H. *(139)*   Dept. of Physcis, Kyoto Univ., Kyoto 606, Japan

Friederich, A. *(32)* Thomson-CSF-Laboratoire Central de Recherches,
B.P.N°10-91401 Orsay, France

Fritzsche, H. James Franck Institute, Univ. of Chicago, 5640 Ellis
Ave., Chicago, IL 60637, U.S.A.

Fujimura, I. *(181)* Ricoh Co., Ltd., 146-1 Matsuhashi Nishi-Sawada,
Numazu, Shizuoka 410, Japan

Fujita, N. *(40)* Sumitomo Electric Ind., Ltd., 1 Miyahigashi Koya,
Itami, Hyogo 664, Japan

Fukunaga, T. *(96)* Dept. of Physics, Osaka Univ., Toyonaka, Osaka 560,
Japan

Fuse, M. *(112)* Ebina Research Lab., Fuji Xerox Co., Ebina, Kanagawa
243-04, Japan

Goda, M. *(153)* Faculty of Engineering, Niigata Univ., Niigata 950-21,
Japan

Hama, T. *(26)* Dept. of Physics, Kyoto Univ., Kyoto 606, Japan

Hamakawa, Y. Faculty of Engineering Science, Osaka Univ., Toyonaka,
Osaka 560, Japan

Hamanaka, H. *(93)* College of Engineering, Hosei Univ., Koganei, Tokyo
184, Japan

Hamano, T. Ebina Research Lab., Fuji Xerox Co., Ebina, Kanagawa
243-04, Japan

Hamasaki, T. *(119)* Dept. of Physical Electronics, Faculty of Engineering,
Hiroshima Univ., Hiroshima 730, Japan

Harada, Y. *(180)* Dept. of Chemistry, College of General Education,
Univ. of Tokyo, Komaba, Megro-ku, Tokyo 153, Japan

Haruki, H. Fuji Electric Corporate Research & Development Ltd.,
2-2-1 Nagasaka, Yokosuka 240-01, Japan

Hatanaka, K. *(130)* Canon Research Center, Canon Inc., 2-2-1 Nakane,
Meguro-ku, Tokyo 152, Japan

Hikami S. *(20)* Research Inst. for Fundamental Physics, Kyoto Univ.,
Kyoto 606, Japan

Hirabayashi, H. *(58)* Toa Nenroy Kogyo K.K., Palace Side Bldg., 1-1-1
Hitotsubashi, Chiyoda-ku, Tokyo 100, Japan

Hirabayashi, I. *(117)* Inst. for Solid State Physics, Univ. of Tokyo,
Roppongi, Minato-ku, Tokyo 106, Japan

Hirai, Y. *(131)* Canon Research Center, Canon Inc., 2-2-1 Nakane,
Meguro-ku, Tokyo 152, Japan

Hiraki, A. *(13)* Faculty of Engineering, Osaka Univ., Toyonaka,
Osaka 560, Japan

172

Kawaguchi, Y. *(21)*   Faculty of Science, Gakushuin Univ., Mejiro, Toshima-ku, Tokyo 171, Japan

Kawamura, H. *(6)*   Dept. of Physics, Kwansei Gakuin Univ., Uegahara, Nishinomiya 662, Japan

Kawaziri, K. *(124)*   Research Lab., Tokyo, Fuji Photo Film Co., Ltd., Asaka, Saitama 351, Japan

Kikuchi, Makoto   Research Center, Sony Corp., 174 Fujitsuka-cho, Hodogaya-ku, Yokohama 240, Japan

Kikuchi, Minoru *(56)*   Research & Development Center, Toshiba Corp., 1 Toshiba-cho, Komukai, Saiwai-ku, Kawasaki 210, Japan

Kim, D.-J. *(55)*   College of Science and Engineering, Aoyama Gakuin Univ., Chitosedai, Setagaya-ku, Tokyo 157, Japan

Kim, G. I. *(146)*   Dept. of Electrical Engineering, Faculty of Engineering, Osaka Univ., Suita, Osaka 565, Japan

Kimura, Y. *(167)*   Dept. of Physics, Tokai Univ., Hiratsuka, Kanagawa 259-12, Japan

Kitao, M. *(88)*   Research Inst. of Electronics, Shizuoka Univ., 3-5-1 Johoku, Hamamatsu, Shizuoka 432, Japan

Kiyono, S. *(178)*   Faculty of Engineering, Tohoku Univ., Sendai 980, Japan

Kobayashi, M.   Dept. of Electronics, Faculty of Engineering, Kyoto Univ., Kyoto 606, Japan

Kondo, M.   Faculty of Engineering Science, Osaka Univ., Toyonaka, Osaka 560, Japan

Kubo, R. *(11)*   Research Inst. for Fundamental Physics, Kyoto Univ., Kyoto 606, Japan

Kuboi, O. *(125)*   Komatsu Electric Metals Co. Ltd., 2614 Shinomiya, Hiratsuka, Kanagawa 254, Japan

Kumeda, M. *(100)*   Faculty of Technology, Kanazawa Univ., 2 Kodatsuno, Kanazawa 920, Japan

Kurata, H. *(123)*   Dept. of Physical Electronics, Faculty of Engineering, Hiroshima Univ., Hiroshima 730, Japan

Kurihara, H. *(61)*   Suwa Seikosha Co., Ltd., 3-3-5 Owa, Suwa, Nagano 392, Japan

Kurobe, A. *(138)*   Dept. of Physics, Univ. of Tokyo, Hongo, Bunkyo-ku, Tokyo 113, Japan

Kuruma, M. *(157)*   Dept. of Electrical Engineering, Gifu Univ., Kakamihara, Gifu 504, Japan

Kuwano, Y. *(85)*   Research Center, Sanyo Electric Co., Ltd., Hashiridani, Hirakata, Osaka 573, Japan

Nakamura, G. *(163)*   LSI Research and Development Lab., Mitsubishi Electric
Corp., 4-1 Mizuhara, Itami, Hyogo 664, Japan

Nakamura, Y. *(44)*   Dept. of Chemistry, Faculty of Science, Hokkaido
Univ., Sapporo 060, Japan

Nakanishi, H. *(164)* Dept. of Physics, Kyoto Univ., Kyoto 606, Japan

Nakashita, T. *(90)*   Dept. of Electrical Engineering, Faculty of Engineering,
Hiroshima Univ., Hiroshima 730, Japan

Nakazato, K. *(162)*   Faculty of Liberal Arts, Shizuoka Univ., 836 Ohya,
Shizuoka 422, Japan

Natori, A. *(87)*   Dept. of Physics, Univ. of Tokyo, Hongo, Bunkyo-ku,
Tokyo 113, Japan

Nemanich, R.J. *(145)* Palo Alto Research Center, Xerox Corp., 3333 Coyote
Hill Rd., Palo Alto, CA 94304, U.S.A.

Ninomiya, T. *(74)*   Dept. of Physics, Univ. of Tokyo, Hongo, Bunkyo-ku,
Tokyo 113, Japan

Nishida, M. *(176)*   Dept. of Electronics, Kanazawa Inst. of Technology,
Nonoichi-cho, Kanazawa 921, Japan

Nishiura, M. *(1)*   Fuji Electric Corporate Research & Development Ltd.,
2-2-1 Nagasaka, Yokosuka 240-01, Japan

Nitta, S. *(62)*   Dept. of Electrical Engineering, Faculty of Engineering,
Gifu Univ., Kakamihara, Gifu 504, Japan

Noda, Mikio *(107)*   Dept. of Technology, Aichi Univ. of Education, Igaya,
Kariya, Aichi 448, Japan

Noda, Mitsunari *(95)* Dept. of Electrical Engineering, Gifu Univ.,
Kakamihara, Gifu 504, Japan

Noguchi, T. *(75)*   Solid State Systems, Development Dept., Semiconductor
Div., Sony Corp., Atsugi Plant, 4-14-1 Asahi-cho, Atsugi,
Kanagawa 243, Japan

Nonomura, S. *(116)*   Faculty of Engineering Science, Osaka Univ., Toyonaka,
Osaka 560, Japan

Oda, T. *(14)*   Imaging Science Engineering Lab., Tokyo Inst. of
Technology, 4259 Nagatsuta, Midori-ku, Yokohama 227,
Japan

Ogawa, Taeko   Dept. of Physics, Kyoto Univ., Kyoto 606, Japan

Ogawa, Tohru   Inst. of Applied Physics, Univ. of Tsukuba, Ibaraki
305, Japan

Oguro, Y. *(50)*   Dept. of Electrical Engineering, Technological Univ.
of Nagaoka, 1603-1 Kamitomioka Nagamine, Nagaoka,
Niigata 949-54, Japan

Ohki, S. *(51)*   Dept. of Electrical Engineering, Technological Univ.
                         of Nagaoka, 1603-1 Kamitomioka Nagamine, Nagaoka,
                         Niigata 949-54, Japan

Ohmi, K.   Fuji Xerox Co., Ltd., Tsukahara-ryo, 3555 Tsukahara,
                         Minami-Ashigara, Kanagawa 250-01, Japan

Ohta, Takeo *(47)*   Material Research Lab. Section MRA1, Matsushita
                         Electric Ind., Co., Ltd., Yakumonaka-machi, Moriguchi,
                         Osaka 570, Japan

Ohta, Tatsuo   Research & Development Lab., Konishiroku Photo Ind.,
                         Co., Ltd., 1 Sakura-machi, Hino, Tokyo 191, Japan

Okamoto, H. *(19)*   Faculty of Engineering Science, Osaka Univ., Toyonaka,
                         Osaka 560, Japan

Okayasu, Y. *(63)*   Toa Nenryo Kogyo K.K., Palace Side Bldg., 1-1-1
                         Hitotsubashi, Chiyoda-ku, Tokyo 100, Japan

Okazaki, M. *(79)*   Inst. of Materials Science, Univ. of Tsukuba, Ibaraki
                         305, Japan

Okuda, M. *(148)*   Dept. of Electronics, College of Engineering, Univ. of
                         Osaka Prefecture, Mozu, Sakai, Osaka 591, Japan

Okushi, H. *(165)*   Electrotechnical Lab., 1-1-4 Umezono, Sakura-mura,
                         Niihari-gun, Ibaraki 305, Japan

Onari, S. *(37)*   Inst. of Applied Physics, Univ. of Tsukuba, Ibaraki
                         305, Japan

Ooeda, H. *(160)*   Electrotechnical Lab., 1-1-4 Umezono, Sakura-mura,
                         Niihari-gun, Ibaraki 305, Japan

Osaka, Y. *(118)*   Dept. of Electronics, Faculty of Engineering, Hiroshima
                         Univ., Hiroshima 730, Japan

Ovshinsky, S.R.   Energy Conversion Devices, Inc., 1675 West Maple Rd.,
                         Troy, MI 48084, U.S.A.

Oyanagi, H. *(102)*   Electrotechnical Lab., 1-1-4 Umezono, Sakura-mura,
                         Niihari-gun, Ibaraki 305, Japan

Ozaki, H.   Dept. of Electrical Engineering, Waseda Univ., 3-4-1
                         Ohkubo, Shinjuku-ku, Tokyo 160, Japan

Paul, W. *(114)*   Div. of Applied Sciences, Harvard Univ., Pierce Hall,
                         Cambridge, MA 02138, U.S.A.

Pietruszko, S.M. *(27)* Solid State Electronics Group, Tata Inst. of Funda-
                         mental Research, Homi Bhabha Rd., Bombay 400005, India
                         (Dept. of Electronics, Inst. of Electron Technology,
                         Technical Univ. of Warsaw, Koszykowa 75,00-662 Warszawa,
                         Poland (permanent address))

Solomon, I. *(103)*   Lab. de Physique de la Matiere, Ecole Polytechnique, Condensee, 91128 Palaiseau Cedex, France

Spear, W.E. *(168)*   Carnegie Lab. of Physics, Univ. of Dundee, Dundee, DD1 4HN, U.K.

Suzuki, Katuhisa *(173)*   Dept. of Physics, Osaka Univ., Toyonaka, Osaka 560, Japan

Suzuki, Kenji *(23)*   Research Inst. for Iron, Steel and Other Metals, Tohoku Univ., Sendai 980, Japan

Suzuki, M. *(30)*   Dept. of Physics, Univ. of Tokyo, Hongo, Bunkyo-ku Tokyo 113, Japan

Suzuki, T. *(92)*   Dept. of Electronics, Hiroshima Inst. of Technology, Itsukaichi, Hiroshima 738, Japan

Takagi, N. *(136)*   Fujitsu Lab. Ltd., 1015 Kamiodanaka, Nakahara-ku, Kawasaki 211, Japan

Takahashi, T. *(17)*   Dept. of Chemistry, College of General Education, Univ. of Tokyo, Komaba, Meguro-ku, Tokyo 153, Japan

Takahashi, Y. *(115)* Dept. of Electric Engineering, Gifu Univ., Kakamihara, Gifu 504, Japan

Takano, Y. *(170)*   Dept. of Electrical Engineering, Waseda Univ., 3-4-1 Ohkubo, Shinjuku-ku, Tokyo 160, Japan

Tanaka, Kazunobu *(34)*   Electrotechnical Lab., 1-1.4 Umezono, Sakura-mura, Niihari-gun, Ibaraki 305, Japan

Tanaka, Keiji *(45)*   Dept. of Applied Physics, Faculty of Engineering, Hokkaido Univ., Sapporo 060, Japan

Tanaka, Y.   Dept. of Physics, Osaka Univ., Toyonaka, Osaka 560, Japan

Tawada, Y. *(18)*   Faculty of Engineering Science, Osaka Univ., Toyonaka, Osaka 560, Japan

Togei, R.   Kawasaki Works, Fujitsu Ltd., Kamiodanaka, Nakahara-ku, Kawasaki 211, Japan

Tokunaga, A.   Dept. of Physics, Osaka Industry Univ., Daito, Osaka 574, Japan

Tsai, C.-C.   Palo Alto Research Center, Xerox Corp., 3333 Coyote Hill Rd., Palo Alto, CA 94304, U.S.A.

Tsu, R. *(5)*   Energy Conv. Dev. Inc., 1675 W. Maple Rd., Troy, MI 48084, U.S.A.

Tsuda, I. *(104)*   2-33-16 Matsugaoka, Takatsuki, Osaka 569, Japan

Tsugane, S. *(120)*   Tokyo National Technical College, Hachioji, Tokyo 193, Japan

Tsuji, K. *(54)*   Inst. for Solid State Physics, Univ. of Tokyo, Roppongi, Minato-ku, Tokyo 106, Japan

Tsujino, K.   Dept of Physics, Osaka Industry Univ. Daito, Osaka 574, Japan

Tsukamoto, J. *(129)* Basic Research Lab., Toray Ind., Inc., 1111 Tebiro, Kamakura 248, Japan

Tsumura, T.   Toa Nenryo Kogyo, K.K., Palace Side Bldg., 1-1-1 Hitotsubashi, Chiyoda-ku, Tokyo 100, Japan

Uchida, Y. *(28)*   Fuji Electric Corporate Research & Development Ltd., 2-2-1 Nagasaka, Yokosuka 240-01, Japan

Usui, S. *(82)*   Research Center, Sony Corp., 174 Fujitsuka-cho, Hodogaya-ku, Yokohama 240, Japan

Wakaki, M. *(101)*   Inst. for Solid State Physics, Univ. of Tokyo, Roppongi, Minato-ku, Tokyo 106, Japan

Wallace, P.R.   Physics Dept., McGill Univ., 3600 University Avenue, Montreal, PQ, Canada H3A 2T8

Watanabe, H. *(91)*   Sendai Radio Technical College, Kami-Ayashi, Miyagi 989-31, Japan

Watanabe, I. *(127)* Faculty of Technology, Kanazawa Univ., 2 Kodatsuno, Kanazawa 920, Japan

Weaire, D. *(86)*   Physics Dept., Univ. College, Belfield, Stillorgen Rd., Dublin 4, U.K.

Weinstein, B.A.   Xerox Webster Research Center 800 Phillips Rd., Bldg., 114 Webster, NY 14580, USA

Yabe, M. *(38)*   Fuji Electric Corporate Research & Development, Ltd., 2-2-1 Nagasaka, Yokosuka 240-01, Japan

Yamabe, T. *(48)*   Dept. of Hydrocarbon Chemistry, Faculty of Engineering, Kyoto Univ., Kyoto 606, Japan

Yamada, K. *(99)*   Lab. of Nuclear Science, Tohoku Univ., 1-2-1 Mikamimine, Sendai, Miyagi 982, Japan

Yamaguchi, M. *(94)* Inst. for Solid State Physics, Univ. of Tokyo, Roppongi, Minato-ku, Tokyo 106, Japan

Yamaguchi, T. *(41)* Faculty of Engineering Science, Osaka Univ., Toyonaka, Osaka 560, Japan

Yamamoto, H. *(25)*   Toa Nenryo Kogyo, K.K., Palace Side Bldg., 1-1-1 Hitotsubashi, Chiyoda-ku, Tokyo 100, Japan

Yamamoto, M.   Dept. of Physics, Osaka Industry Univ., Daito, Osaka 574, Japan

Yamamoto, T. *(126)*   Dept. of Physical Electronics, Faculty of Engineering, Hiroshima Univ., Hiroshima 730, Japan

Yamamoto, Y. *(144)*   Central Research Lab., Engineering Center, Sharp Corp., 2613-1 Ichinomoto-cho, Tenri, Nara 632, Japan

Yamasaki, S. *(174)*   Electrotechnical Lab., 1-1-4 Umezono, Sakura-mura, Niihari-gun, Ibaraki 305, Japan

Yamauchi, Y. *(140)*   Central Research Lab., Engineering Center, Sharp Corp., 2613-1 Ichinomoto-cho, Tenri, Nara 632, Japan

Yano, M. *(171)*   Central Research Inst., Teijin Ltd., 4-3-2 Asahigaoka, Hino, Tokyo 191, Japan

Yao, M. *(134)*   Dept. of Physics, Kyoto Univ., Kyoto 606, Japan

Yasui, K. *(49)*   Dept. of Electrical Engineering, Technological Univ. of Nagaoka, 1603-1 Kamitomioka Nagamine, Nagaoka, Niigata 949-54, Japan

Yokota, M. *(68)*   Faculty of Engineering, Osaka City Univ., 459 Sugimoto-cho, Sumiyoshi-ku, Osaka 558, Japan

Yonezawa, F. *(8)*   Research Inst. for Fundamental Physics, Kyoto Univ., Kyoto 606, Japan

Yoshioka, Y.   Central Research Lab., Sharp Corp., 2613-1 Ichinomoto-machi, Tenri, Nara 632, Japan

Yoshida, A. *(57)*   Dept. of Electric Engineering, Nagoya Univ., Chikusa-ku, Nagoya 464, Japan

Yukimoto, Y. *(15)*   LSI Research and Development Lab., Mitsubishi Electric Corp., 4-1 Mizuhara, Itami, Hyogo 664, Japan

Amorphous Semiconductors

Editor: M. H. Brodsky
1979. 181 figures, 5 tables. XVI, 337 pages
(Topics in Applied Physics, Volume 36) ISBN 3-540-09496-2

Contents:
M. H. Brodsky: Introduction. – *B. Kramer, D. Weaire:* Theory of
Electronic States in Amorphous Semiconductors. –
E. A. Davis: States in the Gap and Defects in Amorphous
Semiconductors. – *G. A. N. Connell:* Optical Properties of
Amorphous Semiconductors. – *P. Nagels:* Electronic Trans-
port in Amorphous Semiconductors. – *R. Fischer:* Lumi-
nescence in Amorphous Semiconductors. – *I. Solomon:* Spin
Effects in Amorphous Semiconductors. – *G. Lucovsky,
T. M. Hayes:* Short-Range Order in Amorphous Semicon-
ductors. – *P. G. LeComber, W. E. Spear:* Doped Amorphous
Semiconductors. – *D. E. Carlson, C. R. Wronski:* Amorphous
Silicon Solar Cells.

Solar Energy Conversion
Solid-State Physics Aspects

Editor: B. O. Seraphin
1979. 209 figures, 13 tables. XIII, 336 pages
(Topics in Applied Physics, Volume 31) ISBN 3-540-09224-2

Contents:
B. O. Seraphin: Introduction. – *B. O. Seraphin:* Spectrally
Selective Surfaces and Their Impact on Photothermal Solar
Energy Conversion. – *A. J. Sievers:* Spectral Selectivity of
Composite Materials. – *H. Gerischer:* Solar Photoelectrolysis
with Semiconductor Electrodes. – *K. Graff, H. Fischer:* Carrier
Lifetime in Silicon and Its Impact on Solar Cell Charac-
teristics. – *M. Savelli, J. Bougnot,* (with the collaboration of
F. Guastavino, J. Marucchi, H. Luquet): Problems of the
Cu_2/CdS Cell. – *A. L. Fahrenbruch, J. Aranovich:* Heterojunc-
tion Phenomena and Interfacial Defects in Photovoltaic Con-
verters. – References. – Several Papers on Solar Energy
Physics Published in **Applied Physics.** – Additional References with Titles. – Subject Index.

Glassy Metals I
Ionic Structure, Electronic Transport, and Cristallization

Editors: H. Beck, H.-J. Güntherodt
1981. 120 figures, 12 tables. Approx. 280 pages
(Topics in Applied Physics, Volume 46) ISBN 3-540-10440-2

Contents:
H. Beck, H.-J. Güntherodt: Introduction. – *P. Duwez:* Metallic
Glasses-Historical Background. – *T. Egami:* Structural Study
by Energy Dispersive X-Ray Diffraction. – *J. Wong:* Exafs
Studies of Metallic Glasses. – *A. P. Malozemoff:* Brillouin Light
Scattering from Metallic Glasses. – *J. Hafner:* Theory of the
Structure, Stability, and Dynamics of Simple-Metal Glasses. –
P. J. Cote, L. V. Meisel: Electrical Transport in Glassy Metals. –
J. L. Black: Low-Energy Excitations in Metallic Glasses. –
W. L. Johnson: Superconductivity in Metallic Glasses. –
U. Herold, U. Köster: Crystallization of Metallic Glasses.

Springer-Verlag
Berlin
Heidelberg
New York

Amorphous Solids

Low Temperature Properties

Editor: W. A. Phillips
1981. 72 figures, approx. 1 table.
Approx. 220 pages
(Topics in Current Physics, Volume 24)
ISBN 3-540-10330-9

Contents:
W. A. Phillips: Introduction. – *D. Weaire:* The
Vibrational Density of States of Amorphous
Semiconductors. – *R. O. Pohl:* Low Tempera-
ture Specific Heat of Glasses. – *W. A. Phillips:*
The Thermal Expansion of Glasses. –
A. C. Anderson: Thermal Conductivity. –
S. Hunklinger, M. v. Schickfus: Acoustic and Di-
electric Properties of Glasses at Low Tempe-
ratures. – *B. Golding, J. E. Graebner:* Relaxa-
tion Times of Tunneling Systems in Glasses. –
J. Jäckle: Low-Frequency Raman Scattering in
Glasses.

Excitons

Editor: K. Cho
1979. 118 figures, 8 tables. XI, 274 pages
(Topics in Current Physics, Volume 14)
ISBN 3-540-09567-5

Contents:
K. Cho: Introduction. – *K. Cho:* Internal Struc-
ture of Excitons. – *P. J. Dean, D. C. Herbert:*
Bound Excitons in Semiconductors. –
B. Fischer, J. Lagois: Surface Exciton Polari-
tons. – *P. Y. Yu:* Study of Excitons and Exciton-
Phonon Interactions by Resonant Raman and
Brillouin Spectroscopies.

O. Madelung

Introduction to Solid-State Theory

Translated from the German by B. C. Taylor
1978. 144 figures. XI, 486 pages
(Springer Series in Solid-State Sciences,
Volume 2)
ISBN 3-540-08516-5

Contents:
Fundamentals. – The One-Electron Approxi-
mation. – Elementary Excitations. – Electron-
Phonon Interaction: Transport Phenomena. –
Electron-Electron Interaction by Exchange of
Virtual Phonons: Superconductivity. – Inter-
action with Photons: Optics. – Phonon-
Phonon Interaction: Thermal Properties. –
Local Description of Solid-State Properties. –
Localized States. – Disorder. – Appendix: The
Occupation Number Representation.

The Physics of Selenium and Tellurium

Proceedings of the International Conference
on the Physics of Selenium and Tellurium
Königstein, Fed. Rep. of Germany,
May 28–31, 1979
Editors: E. Gerlach, P. Grosse
1979. 210 figures, 22 tables. X, 281 pages
(Springer Series in Solid-State Sciences,
Volume 13)
ISBN 3-540-09692-2

Contents:
Bands and Bonds in Se and Te. – Lattice
Dynamics of Trigonal Se and Te. – Bandstruc-
ture in the Neighbourhood of the Gap of
Trigonal Se and Te. – Inperfections and
Impurities in Te. – Transport Phenomena in
Trigonal Se and Te. – The Amorphous,
Glassy, and Liquid State. – Photoelectric and
Transport Phenomena in Amorphous
Systems. – Crystalline and Amorphous
As_2Se_3. – Preparation and Application. –
Index of Contributors.

Springer–Verlag Berlin Heidelberg New York